D1667828

Katrin Scheibe, Franziska Zimmer
Asylees' ICT and Digital Media Usage

Knowledge and Information

Studies in Information Science

Edited by
Wolfgang G. Stock

Katrin Scheibe, Franziska Zimmer

Asylees' ICT and Digital Media Usage

New Life – New Information?

DE GRUYTER
SAUR

ISBN 978-3-11-067192-6
e-ISBN (PDF) 978-3-11-067202-2
e-ISBN (EPUB) 978-3-11-067209-1
ISSN 1868-842X

Library of Congress Control Number: 2021948907

Bibliographic information published by the Deutsche Nationalbibliothek
The Deutsche Nationalbibliothek lists this publication in the Deutsche Nationalbibliografie; detailed bibliographic data are available on the Internet at http://dnb.dnb.de.

Typesetting: Integra Software Services Pvt. Ltd.
Printing and binding: CPI books GmbH, Leck

www.degruyter.com

Acknowledgements

Throughout the writing process of this monograph we have received great support and assistance from numerous people.

First and foremost, we would like to thank Prof. Wolfgang G. Stock, the one who gave us the opportunity to publish this book, the person whose expertise we owe our academic knowledge to, and who pushes us to accomplish the best we can. He was head of the Department of Information Science at Heinrich Heine University Düsseldorf in Germany, is our supervisor, and a good friend. He supported us during the interviewing process, during the writing process, and furthermore supported us in our work. We express our sincere gratitude to him. We would also like to thank his caring and loving wife, Mechtild Stock, who is a beautiful soul and always heartily welcomes us to their home.

Also, we would like to express gratitude towards our colleague Mohamed Nasser Abdillah Abdullh for translating the German survey into Arabic and for his assistance and translation during the interview process. He further encouraged and motivated us with his kind words towards our effort. Additionally, we would like to acknowledge the help of our colleague Dr. Aylin Imeri for proofreading some parts of the manuscript and also would like to thank her and further colleagues, namely Dr. Isabelle Dorsch, Dr. Kaja J. Fietkiewicz, Dr. Maria Henkel, Maurice Schleußinger, Dr. Tobias Siebenlist, and Anneliese Volkmar, for mental support during the writing process and during the COVID-19 pandemic.

Special acknowledgements are further addressed to the interviewed experts, Prof. Dr. Rianne Dekker, Dr. Juliane Stiller, Dr. Violeta Trkulja, and Prof. Dr. Carola Richter. Thank you very much for your time, answering our questions, and sharing your experiences with us. It was very nice to meet you, unfortunately only virtually due to the COVID-19 pandemic.

Further acknowledgements go to the several organizations and institutions which are not mentioned because of data privacy reasons, and further people who supported us:

- Rasia Haji and Eaman Haji for connecting us with children of asylum seekers and organizing interviews.
- Mr. Gerhard Sußek from the Bundesamt für Migration und Flüchtlinge for sharing potential contact addresses of German language courses for asylum seekers.
- Mrs. Elke Faust, Mrs. Silvia Gómez, and Mr. Bernd Sauerwein-Fox for welcoming us with their loving and caring attitude and the opportunity to interview the attendees of German language classes. Furthermore, our acknowledgements go to the German language class teachers.

https://doi.org/10.1515/9783110672022-202

– Mr. Shekib Akbary who reported about his personal experiences as a German language class teacher for asylum seekers.

Lastly, we would also like to thank the refugees and asylum seekers who took part in our interviews and those who answered the survey. We wish them all the best.

In addition, each of the authors would like to personally thank some people:

Special thanks from Katrin Scheibe:

First and foremost, I would like to thank the most important persons in my life, my parents Olga Scheibe and Klaus Scheibe. I am more than grateful to have them in my life. They are always by my side, do their best for me, and support me in every decision of my life. I can always count on them at any time of the day. Further, I would like to acknowledge the patience and love of my partner Tobias Kinle. He always stands by me and supports me. I am thankful to have him in my life. Also, I would like to thank all other members of my family, the members of my partner's family, and my godparents (Andrea and Simon) for supporting me. Furthermore, I am thankful for my loving and caring neighbours and friends, Ramona and Julius, Julia and Patrick, Stefan, and Chantal.

Finally, my special thanks go to my co-author, colleague, and friend, Franziska Zimmer, for her endurance, perseverance, and patience; for staying up all night on some days to finish a research paper in time before a deadline; for taking the journey to Ireland together by train and ferry and visiting the European Google Headquarter office in Dublin; for spending our break between interviews and surveys on Valentine's Day together at a cafe and treating ourselves. Thank you for walking this path together.

Special thanks from Franziska Zimmer:

I like to thank the most important people, my family and friends, Papa, Laura, Jacky, Sarah, Philipp, Christian, Nico, and of course Nine, who support me in more ways than they realize, thank you I love you.

I am grateful to have this first experience together with Katrin. With her creativity, dedication, and immaculate attention to detail (to find even the tiniest error) this book and many more research endeavors were made possible. Thank you for encouraging each other during every phase of this process. And thank you for visiting the Harry Potter Studio Tour together.

Contents

1 Introduction

Difficult conditions, be it war, threat of persecution, or economic and other political reasons, make it necessary for some people to flee their home country and search for asylum. Their situation, living conditions, and circumstances changed. Some are separated from their family members, have to establish new social contacts, and try to stay in touch with their acquaintances and relatives. They have to learn a new language to cope with everyday life and search for job opportunities. Additionally, asylees are faced with the difficult tasks of understanding the bureaucratic requirements of the target country and of cooperating with the administrations. As the asylum applicants need to build a new life, they have to adapt to new situations. They have to reorient themselves to new individual, social, and economic environments. Looking at those changes and adaptations from an information science point (Stock & Stock, 2013): What needs do they have in the new country, what information do they require? Which information services, social media, and information and communications technology (ICT) devices do they apply in order to satisfy their needs? What gratifications do they seek? What is their information behavior and how did their information behavior change?

During the 2015 European migrant crisis, 1.2 million first-time asylum applications were made in the European Union, double than in 2014. Most of these appeals were made by Syrians, of which 50% were filed in Germany. In 2020, there were still around 400,000 applications; it is estimated that the number would be higher since the COVID-19 outbreak and resulting border closures and restrictions of movement prevent new arrivals. Due to political changes in Afghanistan in 2021 when the Taliban took over the country, new migrants may arrive in Europe in the near future. A benefit of this study is to explore the information behavior of asylum seekers – including problems of optimal information dissemination – and can counteract failures and shortcomings in the communication with asylum seekers and vice versa of asylees with people and administrations in their new home country. In the empirical part of this research, the focus is on Germany as a target country, but it can be assumed that some of the findings could be similar and generalizable for other Western countries.

As legal aspects are relevant for asylum seekers, a short overview of the German law is given. Before being registered as a refugee, every asylum seeker has to go through a Refugee Status Determination process which each country can handle differently. Several limitations are set on asylum seekers before being granted some form of protection in Germany. For example, they cannot get regular employment and are not even free to move in the country for the first three months.

https://doi.org/10.1515/9783110672022-001

To be able to orient oneself in the new country, according to the Federal Office for Migration and Refugees (*Bundesamt für Migration und Flüchtlinge*), all newly arrived asylum seekers receive official documents in their native language.

After, in some cases, enduring long and tedious journeys, a new life needs to be established. "Since the start of the European 'migration crisis' there has been a greater awareness about the use of ICTs and social media, which has become one of the major characteristics of migration flows in 2015 and 2016" (Frouws et al., 2016, p. 14). Another important aspect in this regard is the information circulated on social media, sometimes even by smugglers. These practices can be dangerous as smugglers downplay risks associated with these journeys. Verifying this type of information can be difficult and make asylum seekers vulnerable to manipulation. However, being able to compare information on social media about journeys that were taken by others can increase the likelihood of undertaking the long journeys themselves (Frouws et al., 2016). Since it was observed that at least some asylum seekers need to have digital literacy skills and showcase various forms of media practices to manage the long journeys, it is interesting to see how these practices adapt to a new environment. Further, not all asylum seekers are adept at managing various forms of media, so it needs to be investigated how different groups interact with media. As we live in a digital world, technology and media are an integral part of people's everyday life.

Studies investigated the use of ICT and media before and during escape; however, a focused analysis of the information behavior and media usage after forced migration is missing. For instance, it is known that smartphones and social media play an important part during forced migration and serve different purposes, such as keeping in contact with friends and family (Gillespie et al., 2016), or using it as a navigation and translation service (Dekker et al., 2018). It has been already investigated how smartphones as well as digital and social media are important aspects for integration of asylum seekers (Emmer et al., 2016). However, how do information practices change, as the circumstances and information horizons need to be adapted to a new environment?

Information and communications technologies (ICTs) play an important part in this as those products and services (e.g., personal computers, smartphones, search engines, social media, messaging services) enable their users to search for, find, understand, and apply information. ICT skills to use digital technologies efficiently are a growing requirement for various occupations in the majority of countries (OECD, 2016).

First, the basic concepts considered for this research need to be defined. According to the International Organization for Migration (IOM) (2019, p. 137), the term *migration* means "[t]he movement of persons away from their place of usual residence, either across an international border or within a State." Therefore, "an international migrant is any person who changes his or her country of usual residence" (United Nations, 1998, p. 9). Whereby "[a] person's country of usual residence is that in which the person lives, that is to say, the country in which the person has a place to live where he or she normally spends the daily period of rest" (United Nations, 1998, p. 9). For the global context, the European Migration Network (EMN) (2018) adds that migration refers to leaving the country for more than one year. In this context, the term *integration* also needs to be defined. According to the EU Common Basic Principles, adopted in 2004, integration is defined as "a dynamic, two-way process of mutual accommodation by all immigrants and residents of EU Member States" (European Migration Network, 2021). Similarly, the 1951 Refugee Convention outlines the rights of displaced people to participate in the social, political, economic, and cultural spheres of their new society and highlights the responsibility of host countries to create the conditions that enable integration (e.g., access to employment and services) and an acceptance of refugees in the host society.

Important terms related to information and communications technology, media, and their subtypes should be considered as well. Information and communications technology (ICT) is the extensional term of information technology (IT). ICTs are used for unified communication, integrating telecommunication (such as telephone lines and wireless signals) and computers. ICT is an umbrella term referencing any communication device, radio, the Internet, television, cell phones or smartphones, computer and network hardware, or satellite systems (Pratt, 2019). Since the development of ICT, society and its behavior changed, face-to-face interactions adapted to communication via digital spaces. This phenomenon is now termed as the *Digital Age* (Pratt, 2019). With this development, media and its formats also changed. Media includes everything that is concerned with a means for communication and used to reach and influence people, such as radio and television, newspapers, magazines, and the Internet. It can be distinguished between *old media* or *traditional media* and *new media*. Traditional media concerns print media such as newspapers and magazines, books, broadcast television, the radio, letters, phone calls for a landline, and any other forms that do not make use of the Internet. New media describes all media that is digital, interactive, hypertextual, virtual, networked, and simulated in nature (Lister et al., 2009). Included in this definition are components of the communications industry such as digital media (Lister et al., 2009). Digital media needs to be considered as a special form of media in the Digital Age. Any media that is machine-readable encoded is considered digital media (Smith, 2013). Or to put it another way, digital media describes any information or media that is broadcasted through a screen (Smith, 2013), including text, audio, video, and graphics transmitted over the Internet (Rayburn, 2012). Other forms of digital media are software, video games, web pages, websites, databases, digital audio, or electronic documents and books. Even traditional media such as newspapers can be considered digital media if they are uploaded to the Internet (Smith, 2013). These media are used to communicate with more than one person, for example, through books, magazines, film, radio, or television (Smith, 2013). With the addition of the Internet, this kind of communication is now made interactive (Smith, 2013), resulting in the development of a new media form: social media. By posting pictures and videos or sending texts, messages and information are received and returned. On social networking services (SNSs) such as Facebook, groups can be formed to exchange information with many people at once (Knautz & Baran, 2016). Through this information exchange, projects and group efforts can be coordinated, communicated, and collaborated on (Smith, 2013).

Literature on media and ICT usage before forced migration shows that, while living in the former home country, smartphones were seen as useful rather than a necessity (Gough & Gough, 2019). However, during war times, social media was valued for its wealth of information. Furthermore, it is possible to check the situation, for example, via WhatsApp group chats (Gough & Gough, 2019). But, asylum seekers are also aware of misinformation circulating on social media (Alencar et al., 2018). Information that is collected is often about the journey's costs, how to apply for asylum, and what life will be like in the new destination (Koikkalainen et al., 2019). Sometimes false propositions would circulate about the desired new home country, painting a divergent picture about desired living conditions (Dekker et al., 2018; Emmer et al., 2016). Because of ICT and media, irregular migration journeys changed (Frouws et al., 2016). By being able to access information and use ICT such as smartphones, people are more self-reliant and independent. An opinion leader would be selected on the journey, often based on his or her skill to accurately use the smartphone and various apps such as navigation or translation apps (Gillespie et al., 2018). Important are messaging services such as Facebook Messenger and WhatsApp to stay in contact with friends and family (Borkert et al., 2018). During migration, important information is related to travel routes, how to satisfy basic needs such as food, security, or housing, prices of taxis, hotels, where to charge phones, and much more (Ullrich, 2017). How do these information needs change when arriving in a new home country? This also translates to information practices when using media – what information do asylum seekers need in a new environment? What other needs do they have concerning ICT and media and what is their information behavior? Information behavior is not a standalone term, but rather an “umbrella concept” (Savolainen, 2007, p. 109) that unifies all activities related to information, including information need, information production, information seeking, and information reception at work but also during everyday life (Wilson, 2000; Spink & Cole, 2004; Case 2007). In context with information behavior the concept of an *information horizon* needs to be considered. Everyone has social contacts and networks as well as certain contexts and situations that determine the information horizon (Sonnenwald, 1999), which has an impact on the human information behavior and information practices of each individual. Different people might have diverse needs and therefore seek various kinds of information. Besides a desire for information, different needs that are relevant to media usage research are described by the Uses and Gratifications Theory (Katz et al., 1973). The theory describes four motives experienced by media users: a need for information (knowledge), entertainment, social interaction, and self-identity or self-presentation (Katz et al., 1973; McQuail, 1983; Shao, 2009). Uses and Gratifications Theory serves as a basis for some of the investigations of this work.

The present work contributes to the question about how the behavior of asylum seekers changes in the new home country and what information needs they express. Details about the European migrant crisis, including numbers and dates about, as well as reports from asylum seekers fleeing their home country are presented (chapter 2). An introduction about the media and ICT use of refugees and asylum seekers before and during escape follows, highlighting the importance and relevance of these factors during forced migration (chapter 3). Furthermore, diverse initiatives including ICT and media practices to help asylum seekers during the integration process are described (chapter 4). For a better understanding of the asylum seekers' situation in the new home country, details about legal aspects in relation to the German asylum procedure and the asylum process are included (chapter 5). Founding our own empirical research and considering the concept of information behavior and the Uses and Gratifications Theory as a theoretical base, an intuitive research model on the information behavior and media usage of asylum seekers is established and elaborated on (chapter 6). Details about the multiple methods approach including a literature review, interviews with asylum seekers, conducted surveys, applied content analysis, and expert interviews are presented (chapter 7). The literature review concentrates on insights about ICT and media practices of asylum seekers while living in a new home country and related difficulties for asylum seekers (chapter 8). Next, the ICT and media usage of asylum seekers in the new home country, their motives to apply media and ICT as well as their information needs are displayed while taking into consideration age- and gender-dependent results (chapter 9). Subsequently, the ICT and media practices as well as motives to apply media and ICT while living in the former homeland and since living in Germany are compared (chapter 10). Results from the content analysis, revealing the information sought by asylum seekers in forums, follow (chapter 11). To sum up, experts report about their insights on the information behavior and ICT and media usage of asylum seekers and the difficulties during the integration procedures (chapter 12). Finally, brief discussions as well as limitations of the applied methods and an outlook on future work are provided to conclude the work at hand (chapter 13).

This book deepens the understanding of asylum seekers' information behavior and their ICT as well as media usage especially after arriving in a new home country. It shows the benefits of information behavior research to analyze migration and the behavior of (forced) migrants. For our theoretical basis, we worked with a combination of information behavior research and the Uses and Gratifications Theory from media and communication science. The practical value of this book lies in understanding the information and communication processes of asylum seekers, their benefits, and furthermore (and more important) the prevailing information needs in the target country, and therefore information that was not communicated by the responsible authorities. Especially information regarding everyday life, legal aspects, and the asylum procedure seem to be important, but difficulties accessing those information were reported. Besides this, although some asylum seekers use smartphones and other ICTs, they assess their digitial literacy skills as better than they actually are. Mainly older people seem to struggle with handling ICTs and using new media. Those issues can easily be targeted for (possible) future flows of refugees and integration into the new country and culture.

References

Alencar, A., Kondova, K., & Riddens, W. (2018). The smartphone as a lifeline: an exploration of refugees' use of mobile communication technologies during their flight. *Media, Culture & Society, 41*(6), 828–844.

Borkert, M., Fisher, K. E., & Yafi, E. (2018). The best, the worst, and the hardest to find: How people, mobiles, and social media connect migrants in(to) Europe. *Social Media + Society, 4*(1), 1–11.

Case, D. O. (2007). *Looking for Information. A survey of research on information seeking, needs, and behavior* (2nd ed.). Elsevier.

Dekker, R., Engbersen, G., Klaver, J., & Vonk., H. (2018). Smart refugees: How Syrian asylum migrants use social media information in migration-decision making. *Social Media + Society 4*(1), 1–11.

Emmer, M., Richter, C. & Kunst, M. (2016). *Flucht 2.0. Mediennutzung durch Flüchtlinge vor, während und nach der Flucht* [Escape 2.0. Media usage by refugees before, during and after their escape]. Freie Universität Berlin. https://www.polsoz.fu-berlin.de/kommwiss/arbeitsstellen/internationale_kommunikation/Media/Flucht-2_0.pdf

European Migration Network. (2018). *Glossar zu Asyl und Migration – Version 5.0* [Glossary for asylum and migration – Version 5.0]. https://www.bamf.de/SharedDocs/Anlagen/DE/EMN/Glossary/emn-glossary2.pdf?__blob=publicationFile&v=6

European Migration Network. (2021). *Integration.* https://ec.europa.eu/home-affairs/what-we-do/networks/european_migration_network/glossary_search/integration_en

Frouws, B., Phillips, M., Hassan, A., & Twigt, M. (2016). *Getting to Europe the 'WhatsApp' way: The use of ICT in contemporary mixed migration flows to Europe.* Regional Mixed Migration Secretariat. https://mixedmigration.org/wp-content/uploads/2018/05/015_getting-to-europe.pdf

Gillespie, M., Lawrence, A., Cheesman, M., Faith, B., Iliadou, E., Issa, A., Osseiran, S., & Skleparis, D. (2016). *Mapping refugee media journeys. Smartphones and social media networks.* The Open University / France Médias Monde. https://www.open.ac.uk/ccig/sites/www.open.ac.uk.ccig/files/Mapping%20Refugee%20Media%20Journeys%2016%20May%20FIN%20MG_0.pdf

Gillespie, M, Osseiran, S., & Cheesman, M. (2018). Syrian refugees and the digital passage to Europe: Smartphone infrastructures and affordances. *Social Media + Society, 4*(1), 1–12.

Gough, H. A., & Gough, K. V. (2019). Disrupted becomings: The role of smartphones in Syrian refugee's physical and existential journeys. *Geoforum, 105*, 89–98.

International Organization for Migration (IOM). (2019). *International Migration Law – Glossary on Migration.* https://publications.iom.int/system/files/pdf/iml_34_glossary.pdf

Katz, E., Blumler, J. G., & Gurevitch, M. (1973). Uses and gratifications research. *The Public Opinion Quarterly, 37*(4), 509–523.

Knautz, K., & Baran, K. S., Eds. (2016). *Facets of Facebook. Use and Users.* De Gruyter Saur.

Koikkalainen, S., Kyle, D., & Nykänen, T. (2019). Imagination, hope and the migrant journey: Iraqi asylum seekers looking for a future in Europe. *International Migration, 58*(4), 54–68.

Lister, M., Dovey, J., Giddings, S., Grant, I., & Kelly, K. (2009). *New media. A critical introduction* (2nd ed.). Routledge.

McQuail, D. (1983). *Mass communication theory.* Sage.

OECD. (2016). *Skills for a digital world.* https://www.oecd.org/els/emp/Skills-for-a-Digital-World.pdf

Pratt, M. K. (2019, July). *ICT (information and communications technology, or technologies).* SearchCIO. https://searchcio.techtarget.com/definition/ICT-information-and-communications-technology-or-technologies

Rayburn, D. (2012). *Streaming and digital media: Understanding the business and technology.* Routledge.

Savolainen, R. (2007). Information behavior and information practice: Reviewing the "Umbrella Concepts" of information-seeking studies. *The Library Quarterly: Information, Community, Policy, 77*(2), 109–132.

Shao, G. (2009). Understanding the appeal of user-generated media: A uses and gratification perspective. *Internet Research, 19*(1), 7–25.

Smith, R. (2013, October 15). *What is digital media?* The Centre for Digital Media. https://thecdm.ca/news/what-is-digital-media

Sonnenwald, D. H. (1999). Evolving perspectives of human information behavior: Contexts, situations, social networks and information horizons. In Wilson, T. & Allen, D. (Eds.), *Exploring the contexts of information behaviour* (pp. 176–190). Taylor Graham.

Spink, A., & Cole, C. (2004). A human information behavior approach to the philosophy of information. *Library Trends, 52*(3), 373–380.
Stock, W. G., & Stock, M. (2013). *Handbook of information science*. DeGruyter.
Ullrich, M. (2017). Media use during escape – A contribution to refugees' collective agency. *spheres: Journal of Digital Cultures*, *4*, 1–11.
United Nations. (1998). *Recommendations on statistics of international migration*. (Series M, No. 58, Rev. 1). https://unstats.un.org/unsd/publication/SeriesM/SeriesM_58rev1e.pdf
Wilson, T. D. (2000). Human information behavior. *Informing science*, *3*(2), 49–56.

2 European migrant crisis

In 2015, a rising number of people fled their home country and arrived in the European Union (EU), often termed *European migrant crisis* (BBC, 2015; Smith-Spark, 2015), or *refugee crisis* (Clayton, 2015; Tomkiw, 2015) by the media. Around 1.2 million first-time asylum applications were made in 2015 in the EU–more than double the amount of 2014 (Table 2.1). The highest number of those applications were filed in Germany (441,800 or 35%), followed by Hungary (174,400; 15%), Sweden (156,100; 12%), Austria (85,500; 7%), Italy (83,200; 7%), and France (70,600; 6%). Most of the displaced people came from the Greater Middle East, additionally from Africa as well as Southeast Europe (Alencar, 2018). The Middle East is constituted of several countries centered in Western Asia, and encompasses Turkey as well as Egypt, with some definitions including the South Caucasus. The largest ethnic groups are Arabs, Turks, Persians, Kurds, and Azeris.

Most of the appeals were made by Syrians (362,800; 29%), of which around 50% were registered in Germany, as well as by Afghans (178,200; 14%), and Iraqis (121,500; 10%) (Eurostat Press Office, 2016). In 2020, there were still about 416,000 first time applications in the EU. According to Eurostat (2021a), due to the COVID-19 outbreak, the number of asylum applications dropped in 2020 as restrictions of movement because of border closures and administrative measures (for example, suspensions of asylum interviews) prevent people from filing for asylum.

Table 2.1: First-time asylum applications in the EU. Source: Eurostat (2021b).

Year	First-time asylum applications
2020	416,950
2019	675,535
2018	602,520
2017	654,620
2016	1,206,055
2015	1,256,580
2014	562,680

The immigrants who arrive in Europe include asylum seekers and economic migrants (UNHCR, 2016). *Asylum seekers* are individuals seeking international protection who fled their home country and cannot return, fearing persecution

https://doi.org/10.1515/9783110672022-002

because of race, religion, or nationality and have not been legally recognized as a refugee. Every *refugee* was initially an asylum seeker but not every asylum seeker will be recognized as a refugee (UNHCR, 2019). An asylum seeker is only permitted to work in the new country if one has been registered as a refugee. The term *economic migrant* describes a person who leaves their home country because of economic reasons. From this point on, if the term refugee or person entitled to protection is used, it is applied in the context of a person being legally accepted in the new country as a refugee or was granted another form of protection according to German law (entitlement to asylum, subsidiary protection, or ban on deportation). An asylum seeker is someone who has not filed an application yet, whereas an asylum applicant is still waiting on a decision if he or she will be granted some form of protection. For this study, the term asylum seeker is used, as the status of the study participants was not determined and not every participant wanted to disclose their status for privacy reasons.

Considering the number of first-time asylum applications divided by gender, there are more men fleeing to the EU than women (Table 2.2). In 2015, it is estimated that 911,060 (72.50%) men and 344,270 (27.39%) women applied for asylum. Men were the ones traveling the often dangerous and difficult journeys; sometimes, a lack of legal routes leave them no other choice than to turn to smugglers. They are also the ones trying to find a safe place to live and work before reuniting with families. This is also evident by the numbers – around half of all asylum seekers want to apply for family reunification once they are settled in a new country (UNHCR, 2015).

Table 2.2: Number of first-time asylum applications in the EU by gender (rounded). Source: Eurostat (2021b), retrieved March 2021.

Year of first-time applications	Male	Female	Unknown	Total
2020	266,175	150,675	100	416,950
2019	422,105	253,235	195	675,535
2018	383,125	219,220	175	602,520
2017	436,985	217,390	240	654,620
2016	814,980	389,150	1,930	1,206,055
2015	911,060	344,270	1,050	1,256,580
2014	398,355	164,155	175	562,680

Looking at the asylum applications by age, for all years, the highest number are the 18–34 year olds (Table 2.3). For example, in 2015, around 668,275 (53.18%) were between 18–34 years old. In 2020, the percentage is similar, 47.69% (198,845) of all applications belong to this age group. The second highest age group for all years is composed of children being younger than 18, with children under 14 as the biggest subgroup. In 2015, around 368,050 (29.29%) of the ones being younger than 18, and of these, 243,215 (19.36%) being younger than 14, filed for asylum in the EU. Similarly, in 2020, 129,560 (31.07%) were younger than 18 and from these, 103,925 (24.92%) younger than 14 who searched for protection in the EU. The older generation (>65) only makes up a very low proportion of the applications (e.g., 0.57% or 7,200 in 2015).

Table 2.3: Number of first-time asylum applications in the EU by age (rounded).
Source: Eurostat (2021b), retrieved March 2021.

Year of first-time applications	Age <14	Age 14–17	Age 18–34	Age 35–64	Age >64	Total
2020	103,925	25,635	198,845	85,280	3,265	416,950
2019	163,095	39,350	320,085	146,395	5,380	675,535
2018	146,445	38,890	291,470	119,765	4,250	602,520
2017	152,955	48,395	336,245	111,430	4,085	654,620
2016	282,525	103,890	616,985	193,820	7,250	1,206,055
2015	243,215	124,835	668,275	211,430	7,200	1,256,580
2014	105,605	38,945	305,435	106,745	4,525	562,680

An overview of all asylum applications in 2015 illustrates the age and gender distribution (Figure 2.1). Males dominate all age groups below the age of 60. This only changes for the people being older than 60. Here, more women applied for asylum.

Overall, as the journeys can be quite tedious, at first, mostly young men are the ones leaving the home country. It can seem like asylum seekers arrive in Europe as a linear and singular uninterrupted flow of people, which is not the case (Crawley et al., 2016). The routes are diverse, many countries are travelled and the time spent in them differs widely. Furthermore, many people were able to travel regularly for parts of the journey, so they did not cross the borders irregularly without a passport or visa. Many decide to catch trains, walk, or pay smugglers to leave their home country for Europe (Deutsche Welle, 2015a). A refugee who fled from Libya describes such a route (UNHCR, 2018, n.p.):

> Where I am from in Somalia, there was extreme drought. People were dying and there was no work. So in 2016, I made the decision to leave. Honestly, I thought it was bad in Somalia . . . but the things I have seen since then on the journey here . . . people being electrocuted, being tortured, people dying even. When I left, I didn't know where I would go, or how I would get there. I took around $1,000 with me, and paid all of it to smugglers in Libya. I went first to Yemen and took a boat for some weeks across to the sea to Sudan. They took all of our luggage and threw it away. On that journey, nobody cares if you live or if you die. The boat was not even big enough to carry us all. When we arrived in Sudan, smugglers took us in a car. There was not enough space in the vehicle. They put people on top of one another. But if you complained, they would beat you. We barely ate for two weeks. We couldn't. Many people were sick and vomiting, two people even died. When we arrived to Libya, we were forced into an underground hole. So many people were sick. My good friend was so sick that he died there . . . You can't sleep in the hole. There are maggots all over you, eating your skin. They give you a phone to call your family to ask for money, and they beat you and electrocute you. When I refused to call my family, they tied my hands and feet, and poured water all over me and electrocuted me . . . they electrocute you until your whole body shakes.

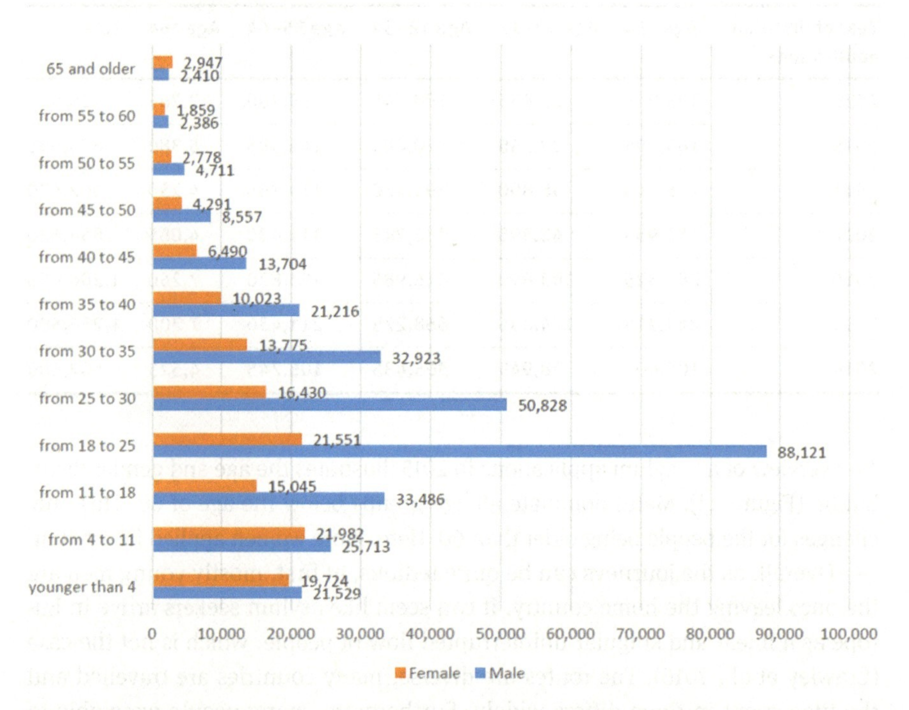

Figure 2.1: Gender and age pyramid of asylum applications in Germany, 2015.
Source: BAMF (2016).

Many people fleeing their home country such as Syria, Iraq, Afghanistan, and Yemen cite the activities of Islamic State of Iraq and the Levant (ISIL, or IS) as the reason to leave (Crawley et al., 2016). Furthermore, kidnapping by a range of different state and non-state actors as a threat for them and their families was named by Syrians and Iraqis. Eritreans, Syrians, and Afghans (living in Iran) also named the fear of being forcibly conscripted into the government army, militia, or rebel force. Institutionalized discrimination on the basis of ethnic or religious identity was stated as another big reason to leave for Hazara Afghans from Iran, Christians from Eritrea, or Palestinians from Syria (Crawley et al., 2016). Another refugee describes (Crawley et al., 2016, p. 10):

> There is nothing left after 5 years of war. Yet in order to leave, I had to go through IS checkpoints. My plan was to go to Lebanon and find a job there. We went by car to another area in Raqqa which was controlled by the regime. We paid a smuggler 50 dollars per person. The car was full of clothes, and we were hidden among these clothes. It was a transportation vehicle supposedly. We first went to Palmyra, and from there to Damascus. We were stopped in Palmyra and asked where I was going and why I was leaving. It wasn't allowed to leave the city if you are younger than 40 years old. In case you are stopped, you must not say that you are going to Lebanon. I told them that we were going to another city controlled by IS. The regime stopped us too. They checked our IDs. They cross check your ID in a computer, and if you are 'clean' they let you go. We arrived in Damascus and we went by coach to the border with Lebanon.
>
> (Syrian man, travelling with wife and four children aged under 11)

When fleeing from Syria or neighboring countries, the first stop is often the province of Izmir in Turkey. There, people sleep in hostels, tents set up in parks, or are forced to sleep on the street (Deutsche Welle, 2015a). Following, the routes spanned across the sea, reaching the Greek island of Kos, often by using inflatable dinghies or other boats. However, Kos was not prepared for the influx of new arriving people, creating difficult living conditions. From Kos, the next step leads to the Greek mainland port of Piraeus by a ten-hour drive via ferry. From the port of Piraeus, the route goes north to the Greek border to Macedonia, the town of Idomeni. Continuing to Serbia, this portion of the journey has since been termed the *Balkan route*. Macedonia also struggled with the sudden influx and declared a state of emergency, mobilizing their army to prevent refugees from crossing its borders. This led to complications, as asylum seekers assumed based on previously lax border controls these routes would be safe. To reach Serbia, crammed trains are boarded. Reportedly, Macedonia and Serbia were only able to offer fewer than 3,000 places of reception collectively. In contrast, there were 19,000 new arrivals in the first weeks of June alone (UNHCR, 2015). Often, refugees had to walk on railway tracks or were walking or cycling among the emergency lane of highways as they were legally omitted from using public transport. Later, this law was changed to protect asylum seekers from serious injuries resulting from those practices (UNHCR, 2015). Hungary is the next destination. Yet, Budapest increased their punishments for illegal border crossings, further complicating the journey. Austria should be reached next, but, due to Budapest's strict rules, refugees were often held back at Bicske Station. Hungary wanted to process the asylum requests that were filed mandatorily, however, many refused to apply as they fear being sent back to Hungary if they are caught later in Western or Northern Europe. After days staying in Hungary, Austria and Germany are reached (Deutsche Welle, 2015a).

Another popular route is the one taken via the Mediterranean Sea (UNHCR, 2015). Accordingly, the number of people arriving in 2015 were of historic proportions. Due to unsafe boats or dinghies, many people lost their lives trying to reach Europe. For example, in 2015 between January and March, 479 drowned or went missing on sea. Syrians made up the largest group of arrivals to Europe, followed by Eritreans and Afghans in 2015. In 2014, most of the arriving refugees reached the Italian shores (around 170,000 in the first three quarters of 2014), followed by Greece (43,500) via the Mediterranean Sea. The number of arrivals changed in 2015: during the first six months, 68,000 arrived in Greece and 67,500 in Italy. The latter one received mainly people from Eritrea (25%), Nigeria (10%), Somalia (10%), Syria (7%), and Gambia (6%). In Greece, the countries of origin differ: Syria (57%), Afghanistan (22%), and Iraq (5%) were most prominent. However, Greece is not seen as the final destination, as 90% want to find asylum somewhere else in the EU, preferably in Germany or Sweden. This also related to the harsh conditions in Greece and Italy faced by asylum seekers as both countries were not prepared for the sudden large influx. For example, concerning Greece, in 2015, almost 20% of asylum seekers did not have access to a toilet and 70% did not regularly receive hygiene items. More than half had no access to a shower or blankets and 30% had no mattress to sleep on (UNHCR, 2015). The journey for most of the asylum seekers arriving in the Southern European countries followed the Balkan route to reach Northern Europe – countries like Sweden and Germany are perceived as being able to offer more protection, better support for asylum seekers, a more welcoming environment, and easier prospects for integration. This is also apparent by the number of asylum applications – Italy saw 67,500 arrivals but only 28,500 asylum applications in the first half of 2015. Greece registered similar numbers, with 68,000 arrivals and 5,115 applications in the same time span (UNHCR, 2015).

A different trend has since emerged in the last few years when it comes to preferred routes (UNHCR, 2018). Italy has seen a decrease in arrivals by 80% from 2017 (119,400) to 2018 (23,400). Now, another country faces large influxes of refugees and became the primary entry point by sea: Spain. The arrivals increased by 131% from 2017 (28,300) to 2018 (65,400). From there, the journey continues to France, often from the Spanish town of Irun to the French town of Bayonne.

Further, the nationalities of arriving individuals changed as well. People are fleeing from Morocco (13,000), Guinea (13,000), and Mali (10,300). The demographics are similar – men make up the largest number with 78%, followed by women and children with 11% respectively. Reasons to leave the home countries are stated as: fleeing gender-related persecution such as forced marriage, persecution on account of sexual orientation or gender identity, and political persecution. Those who arrived in Greece also changed their routes through the Balkans in contrast to previous years. Now, transiting through Bosnia and Herzegovina to other EU member states instead of Serbia to Hungary are preferred since the latter routes became more restrictive (UNHCR, 2018).

Europe was not seen as the final destination at first, but the conditions in other first sought after host countries forced asylum seekers to travel further – stated reasons are a lack of rights, limited employment, health, education, or discrimination. Family reunion and the policies relating to refugee status were more important than access to welfare support in shaping intended destination decisions (Crawley et al., 2016). "I am just looking for peace and a job. I would stay in any place in Europe that could offer me these two things" (Crawley et al., 2016, p. 9). Many people found refuge in Germany, but why did they choose this country as their new home? The main reason is that Germany has since welcomed many asylum seekers, establishing solid networks for migrants. Germany is economically strong and has a positive future outlook on politics, society, and economics (Oltmer, 2016).

North Rhine-Westphalia is the Federal State (*Bundesland*) with the highest number of appeals since 2015. Then, most asylum seekers came from Syria, followed by Afghanistan, Iraq, and Pakistan. In 2021, not much has changed – many of the 56,687 asylum seekers registered from January to April still originate from Syria (54.10%), Afghanistan (8.63%), Iraq (6.71%), and Turkey (2.89%) (BAMF, 2021).

Due to the rising number of asylum seekers reaching Europe, and in this case especially Germany, the media reported on these events from various points of view. With her open stance on immigration policies, the German chancellor Angela Merkel received praise and criticism. She temporarily suspended the *Dublin rule*, which originally states that all asylum seekers need to stay in the first country they came in contact with. By suspending this rule, asylum seekers were no longer forced to return to the first country of contact, resulting in an influx of Syrian asylum seekers into Germany and further, relieving the pressure on Hungary and Greece (Deutsche Welle, 2015b). Barack Obama praised her, "What's happening with respect to her position on refugees here, in Europe, she's on the right side of history on this" (ABC, 2016). The Guardian's analyst Will Hutton states, "Angela Merkel's humane stance on migration is a lesson to us all. . . .

The German leader has stood up to be counted. . . . She wants to keep Germany and Europe open, to welcome legitimate asylum seekers in common humanity, while doing her very best to stop abuse and keep the movement to manageable proportions. Which demands a European-wide response" (Hutton, 2015). Opposing voices were also shared. For example, former French president Nicolas Sarkozy criticized Merkel for her decision, stating this would lead even more people to Europe, and consequently to France (Jaigu et al., 2015). Opposing views also termed the new influx *Islamic invasion*. As countries like Serbia, Turkey, and Greece are safe, if one wants to flee those, it is done for benefits and a house, as stated by other opposing voices (The Jerusalem Post, 2015).

The impact of the European Refugee Crisis on society and history is yet to be determined, even on country level. If changes of the labor market are concerned, this part of integration can be a long-term process. Earlier waves of refugees needed around 20 years to attain employment rates similar to those of the national population. Germany registers data about the educational level of newly arriving asylum seekers. For 2015, the Federal Office for Migration and Refugees recorded this data for around 73% of the asylum seekers. Of those, 18% had a university-level education and 20% had a higher secondary education. Only 7% had no formal education (Dumont et al., 2016). Integration courses can be beneficial for this. In 2015, around 179,000 (63.25%) asylum seekers attended integration courses, of those who were eligible to do so (283,000). It is expected that the number of asylum seekers and refugees will have a socioeconomic and demographic effect on Germany and Europe (Juran & Broer, 2017). Despite the costs for integration and educational aspects, a 0.5% growth in gross domestic product is projected. Further, new arrivals are expected to contribute to Germany's economic performance in a positive way through increases in labor supply as well as a boost in demand and services. However, this will depend on the extent to which refugees can be fully integrated into German society.

References

ABC. (2016, April 24). Barack Obama says Angela Merkel 'on right side of history' over pro-refugee stance. *ABC*. https://www.abc.net.au/news/2016-04-25/merkel-on-right-side-of-history-over-refugee-stance-obama-says/7354418

Alencar, A. (2018). Refugee integration and social media: a local and experiential perspective. *Information, Communication & Society*, *21*(11), 1588–1603.

BAMF. (2016). *Das Bundesamt in Zahlen 2015. Asyl, Migration und Integration* [The federal office in figures 2015. Asylum, migration and integration]. https://www.bamf.de/SharedDocs/Anlagen/DE/Statistik/BundesamtinZahlen/bundesamt-in-zahlen-2015.pdf?__blob=publicationFile&v=16

BAMF. (2021). *Aktuelle Zahlen. Ausgabe April 2021* [Current data. April 2021 edition]. https://www.bamf.de/SharedDocs/Anlagen/DE/Statistik/AsylinZahlen/aktuelle-zahlen-april-2021.pdf?__blob=publicationFile&v=2

BBC. (2015). *Europe migrant crisis*. https://www.bbc.com/news/world-europe-32395181

Clayton, J. (2015, September 4). UNHCR chief issues key guidelines for dealing with Europe's refugee crisis. *United Nations High Commissioner for Refugees*. https://www.unhcr.org/55e9793b6.html

Crawley, H., Duvell, F., Jones, K., & Skleparis, D. (2016). *Research brief 02: Understanding the dynamics of migration to Greece and the EU: Drivers, decisions and destinations*. MEDMIG. http://www.medmig.info/research-brief-02-Understanding-the-dynamics-of-migration-to-Greece-and-the-EU/

Deutsche Welle. (2015a, September 9). A perilous trek: refugees' journey from Syria to Germany. *Deutsche Welle*. https://www.dw.com/en/a-perilous-trek-refugees-journey-from-syria-to-germany/g-18696817

Deutsche Welle. (2015b, August 19). Germany suspends 'Dublin rules' for Syrians. *Deutsche Welle*. https://www.dw.com/en/germany-suspends-dublin-rules-for-syrians/a-18671698

Dumont, J.-C., Liebig, T., Peschner, J., Tanay, F., & Xenogiani, T. (2016). *How are refugees faring on the labour market in Europe? A first evaluation based on the 2014 EU labour force survey ad hoc module* (Working Paper 1/2016). European Union. https://op.europa.eu/en/publication-detail/-/publication/87a8f92d-9aa8-11e6-868c-01aa75ed71a1

Eurostat. (2021a). *Asylum statistics*. https://ec.europa.eu/eurostat/statistics-explained/index.php?title=Asylum_statistics#First-time_applicants:_581_thousand_in_2018

Eurostat. (2021b). Asylum and first time asylum applicants by citizenship, age and sex – annual aggregated data (rounded). *Data Browser*. Retrieved March 15, 2021 from https://ec.europa.eu/eurostat/databrowser/view/MIGR_ASYAPPCTZA__custom_1593290/default/table?lang=en

Eurostat Press Office. (2016, March 4). *Asylum in the EU member states. Record number of over 1.2 million first time asylum seekers registered in 2015*. https://ec.europa.eu/eurostat/documents/2995521/7203832/3-04032016-AP-EN.pdf/790eba01-381c-4163-bcd2-a54959b99ed6

Hutton, W. (2015, August 30). Angela Merkel's humane stance on immigration is a lesson to us all. *The Guardian*. https://www.theguardian.com/commentisfree/2015/aug/30/immigration-asylumseekers-refugees-migrants-angela-merkel

Jaigu, C., de La Grange, A., & Brézet, A. (2015, September 9). Crise des migrants: le plan de Sarkozy [Migrant crisis: Sarkozy's plan]. *Le Figaro*. https://www.lefigaro.fr/politique/2015/09/09/01002-20150909ARTFIG00427-crise-des-migrants-le-plan-de-sarkozy.php

Juran, S., & Broer, P. N. (2017). A profile of Germany's refugee populations. *Population and Development Review*, *43*(1), 149–157.

Oltmer, J. (2016, April 21). Warum ist die Bundesrepublik Deutschland 2015 Ziel umfangreicher globaler Fluchtbewegungen geworden? [Why did the Federal Republic of Germany 2015 become destination for global mass migration?]. *Bundeszentrale für politische Bildung*. https://www.bpb.de/gesellschaft/migration/kurzdossiers/224849/fluchtziel-deutschland

Smith-Spark, L. (2015, September 5). European migrant crisis: A country-by-country glance. *CNN*. https://edition.cnn.com/2015/09/04/europe/migrant-crisis-country-by-country/

The Jerusalem Post. (2015, September 10). Wilders tells Dutch parliament refugee crisis is 'Islamic invasion'. *The Jerusalem Post*. https://www.jpost.com/Middle-East/Wilders-tells-Dutch-parliament-refugee-crisis-is-Islamic-invasion-415828

Tomkiw, L. (2015, September 3). European refugee crisis 2015: Why so many people are fleeing The Middle East and North Africa. *International Business Time*. https://www.ibtimes.com/european-refugee-crisis-2015-why-so-many-people-are-fleeing-middle-east-north-africa-2081454

UNHCR. (2015). *The sea route to Europe: The Mediterranean passage in the age of refugees*. https://www.unhcr.org/5592bd059.pdf

UNHCR. (2016, July 11). UNHCR viewpoint: 'Refugee' or 'migrant' – Which is right?. *United Nations High Commissioner for Refugees*. https://www.unhcr.org/55df0e556.html

UNHCR. (2018). *Desperate journeys*. https://www.unhcr.org/desperatejourneys/

UNHCR. (2019). Refugee status determination. *United Nations High Commissioner for Refugees*. https://www.unhcr.org/pages/4a16b1d06.html

3 ICT and media practices before and during escape

With the rising number of people fleeing their home country and the increased use of technology and social media, governments also became more active on such services. It is argued that social media is used to raise awareness of the dangers associated with irregular migration. This way, the governments were also able to track migration flows (Frouws et al., 2016).

Musarò (2019) analyzed such a campaign conducted by the Italian government to deter potential illegal migrants and asylum seekers from fleeing to Italy. The specific information campaign – *Aware Migrants* (2016) – was funded by the Italian government and managed by the International Organization for Migration (IOM) aimed at preventing people from fleeing via the Mediterranean Sea. Other European countries joined, for example, the German Federal Foreign Office helped to launch later phases of the campaign. It was spread via social media (such as Facebook, YouTube, and Instagram) and a website developed in three languages, English, French, and Arabic. The campaign considered costs and dangers associated with migrant smuggling as well as dangerous and harsh conditions during the journey and after the arrival in Europe. Critical voices about the government campaign state that information is missing on how to use alternative, safe routes for migration:

> [M]ore generally, using fear mongering messages (shock and sorrow) to symbolically connect irregular migration to financial and familial ruin, as well as by emphasizing the illegal aspects of migrant smuggling and labelling smuggled migrants as illegals, this campaign may have the effect of deterring persons in desperate situations, facing prosecution, torture, discrimination and human rights abuses, from exercising their right to seek asylum by keeping them at a distance. (Musarò, 2019, pp. 636–637)

However, there is not enough data on the impact of such campaigns, especially on social media, since they can be based on an inadequate understanding on how asylum seekers source information (Frouws et al., 2016).

3.1 ICT and media practices before forced migration

The following paragraphs deal with the ICT and (social) media usage and information behavior on how asylum seekers search for and apply information before migration. This is also relevant in relation to Musarò's (2019) findings on the inadequate understanding of asylum seekers' information behavior on social media.

https://doi.org/10.1515/9783110672022-003

A study conducted in 2012 about the information gathered by asylum seekers prior to migration shows interesting insights (Fleay et al., 2016). The study participants heard of Australia as a viable option for refuge through Internet sources and friends without conducting any research:

> Through the media we came to know, everyone knows Australia is accepting refugees. [I got information from] television, information available everywhere, you can listen to the radio, you can use the Internet if you like. It is not about getting information, it is about securing a safe place in the world. I didn't do any research. (Fleay et al., 2016, p. 65)

However, misinformation was also spread, sometimes by people-smuggling agents in the home country:

> I didn't have any plan before I came [to Australia]. One of my friends had been in Australia, he had been in Australia from 2000. He told me about Australia, about the lifestyle. I couldn't read in English well, I was not searching and planning [about coming to Australia]. The organizing people in my country, agents, through one of my friends I find out. [The agent] told me a lot of things in Australia. I was told Australian Government will take you to the shopping centre and you can buy anything you want. Especially for refugees, they have deep sympathy. It's a very rich country, they can afford it easy. You will get a package – like after detention, but [the agent] polished it. He lied. (Fleay et al., 2016, p. 66)

Information on Australian policies were not accessed before or during the journey, the Internet was mainly used "for killing time, nothing special" (Fleay et al., 2016, p. 66). Information sources were not used or research done besides trusting friends or smugglers.

However, one may also observe a different information behavior regarding the preperation for the escape. Some insights were made regarding the ICT and media use of asylum seekers for migration decision making. As established by Dekker et al. (2018), and further studies as well, asylum seekers access information through the Internet on their smartphone rather than other ICTs (Gillespie et al., 2016; Emmer et al., 2020). This is also related to problems caused by war and unrest in the home country. For example, due to power cuts, easy access to TV and the Internet is not always present (Gillespie et al., 2016; Dekker et al., 2018). Another problem is a weak Internet signal: "In Syria, there was little internet connection. At home, I had a router to enhance the signal" (Dekker et al., 2018, p. 5). Therefore, news is usually accessed through smartphones. Further, only 44% or 64% (dependent on the investigated study group) of the study participants by Merisalo and Jauhiainen (2019) were able to use the Internet in their home country (in this case, Syria). This is also related to the fact that the second group of participants came from Aleppo or other urban areas, whereas the first group came from rural and less-developed parts of Syria. Additionally, more men than women use the Internet in their home country as well as younger participants more than older ones. If possible, international news channels are watched on TV (Gillespie et al., 2016; Emmer et al., 2020). Other ICTs and media such as the radio are much less popular, and least used are print media (Emmer et al., 2020). To an extent, laptops and home PCs are sometimes used (Dekker et al., 2018). Smartphones were mostly seen as being useful rather than a necessity, to foster social interactions via social media such as WhatsApp or to be used for phone calls:

> Actually in Syria we don't need this much internet or the mobile because you know you are living with your family and your friends. You don't need all these kinds of programmes [apps] and if you want to go from work to another place you know the streets – you know how to move. It is easy . . . people were not thinking too much about the internet so maybe they use the internet only for chatting on WhatsApp but that's it.
>
> (Gough & Gough, 2019, p. 6)

The role of social media in migration decision making was further analyzed by Dekker et al. (2018) and Gough & Gough (2019). The wealth of information that is shared on social media is valued for its timeliness and personal background. It was used to track the constantly changing situation in the home country during war via group chats on WhatsApp and basing the decision to leave on those exchanges (Gough & Gough, 2019). During war, social media such as Facebook is valued for maintaining social capital (Ramadan, 2017). Facebook enhanced social relationships. Syrians adopted Facebook because they believed in its perceived usefulness for enhancing their social relations (e.g., making new friends and maintaining existing ones, keeping in touch with old friends, and finding lost contacts). Through this, information from these online relationships can be shared and accessed.

Untrustworthiness of information is one of the biggest concerns for asylum seekers when using social media and deciding where to migrate. Information is trusted from well-known online communities as well as from existing social ties (Dekker et al., 2018; Emmer et al. 2016):

> Personal experiences that are shared on Facebook are trustworthy. Those people followed the same routes and have gone through the same procedures. These different stories confirm each other . . . . Public information was not always very trustworthy. Private messages are more trustworthy. For example, when I communicated with someone living in the Netherlands, that gave me a good image of what life in the Netherlands looks like.
>
> (Dekker et al., 2018, pp. 7–8)

Needed information are related to the traveled routes to Europe, access to different countries in Europe and which destination to choose (Dekker et al., 2018) as well as the monetary costs of the journey, how to apply for asylum, and what life in the new destination will be like, although to a smaller extent than the first ones (Koikkalainen et al., 2019). To prepare for the journey, information about the new home country as well as experiences of other asylum seekers were searched for by half of the study participants (Emmer et al., 2020). When no existing social ties to Europe were available, information was preferably accessed through official websites of organizations. As an alternative, Google Maps was used to plan different routes (Emmer et al., 2016). Those information practices empower asylum seekers and make them less vulnerable to fraud and misinformation. Specific types of information were identified that were handled the most carefully by asylum seekers – information about the duration of the residence permit application as well as the application for family reunification:

> I heard that in Germany and Denmark, there is a lot of discrimination toward refugees. I also heard that in Sweden the procedures to acquire a residence permit take a long time. People told me that in the Netherlands, these procedures are shorter, that people are hospitable and that English is an important language that is spoken by many.
>
> (Dekker et al., 2018, p. 7)

Additionally, there seemed to circulate different encouraging and discouraging information and rumors on social media among asylum seekers' networks, for example "they will take your money and your phone, you will end up in prison" (Dekker et al., 2018, p. 7). Around 90% of the Syrian and Iraqi study participants from Emmer et al. (2016) heard rumors that they would receive their own house, free social services, and would be able to reunite with the whole family in Germany. Nevertheless, these rumors or misinformation did not influence those who were set on migrating to a specific country. "I heard many rumours about the situation in the Netherlands. For example, the social benefits were bad. This did not influence my decision. I had a goal and for the future of my children, the social benefits did not matter that much" (Dekker et al., 2018, p. 9). Information is perceived as being a rumor especially if it was spread by unknown sources and was publicly available (Dekker et al., 2018). To verify information, four information behavior strategies were identified: checking the sources of information, validating information with trusted social ties, triangulation of online sources, and comparison of information with own experiences.

An interesting aspect was mentioned by Koikkalainen et al. (2019) as well as Emmer et al. (2016). They found that even though information about a new home country is searched for before migration, rather than learning as much as possible about the immigration policies or the political destination, sometimes, an idealized version of the new country is imagined. "But when I go from Turkey, my mind is just about Finland. Because they told me many more things about Finland. I told you about democracy and safety and like this yeah. And they say the people in Finland are very kind. In my mind: just Finland" (Koikkalainen et al., 2019, p. 61). This is often related to false online marketing or word-of-mouth information.

These sources are social media, friends and relatives, fellow countrymen already living in the new home country, or smugglers. Even though the Internet is used to find accurate information before migration, if this is correlated with TV and international media consumption, it leads to a positive colored view of a new home country, which is not necessarily based on facts but rather an imagined ideal (Emmer et al., 2016). Interesting insights were also made about the image of a new country and media usage. Internet usage is associated with more factual knowledge of the target country, whereas the use of traditional media is not. Additionally, even though Internet usage led to a positively biased image of, in this case, Germany, and print media did not, the effect of traditional media consumption is comparatively stronger. The reason why print media might not generate such a positive image as the Internet could be explained by the reporting of protocol news by print media, mostly focusing on disasters and conflicts in contrast to the Internet where international media offerings are available as well as personal stories via social media (Emmer et al., 2020).

3.2 ICT and media practices during forced migration

It is argued that ICT and social media changed the way irregular migration is planned and navigated (Frouws et al., 2016). Through ICT and social media, information on migration routes can be accessed, smugglers contacted, information on the destination country gathered as well as safety and rescue be organized, travel plans adapted flexibly, and dangers avoided. The use of mobile technology helps to be self-reliant and independent. Further, friends can be reached and information inquired about through online social networks (Frouws et al., 2016; Gillespie et al., 2016; Zijlstra & van Liempt, 2017). This digital navigation is essential, especially since there are few legal alternative tools to check if one is following the appropriate travel route (Gillespie et al., 2018). Traditional media or ICT such as the computer are not used for these crucial tasks (Borkert et al., 2018). Here, the smartphone is the most important ICT device.

As it can be risky to carry a mobile phone when caught by authorities while crossing borders, mobile phones are often sold in the homeland (Gillespie et al., 2016) and later bought in a stopover country, additionally with mobile power banks and waterproof cases to keep the phone protected if travel over water is necessary (Kaufmann, 2016). Further, at those stopover destinations, information about different developments at the borders is exchanged, digital maps and apps are downloaded, digital documents prepared (Kaufmann, 2016), and WiFi spots as well as battery charging resources searched for (Borkert et al., 2018; Gillespie et al., 2018; Gough & Gough, 2019). These steps are often taken just to let friends and family know that the destination was reached safely as relatives worry about their whereabouts because of media postings of boats sinking in the sea. Contact with relatives via smartphones is a significant aspect for mental wellbeing during forced migration. Also, being able to see pictures of family and friends on a smartphone is important to counteract psychological pressures (Kaufmann, 2016). Further, pictures are taken during the journey for memory purposes (Alencar et al., 2018). To this end, mostly positive aspects are represented on these images to send back home and upholding a positive appearance to not worry relatives (Gough & Gough, 2019). Important documents such as IDs and diplomas are photographed and saved digitally on the phone, USB-sticks, or online – they are the base for a new livelihood in a new home country (Kaufmann, 2016; Alencar et al., 2018). Also, through ICT, large sums of money do not have to be carried around anymore (Frouws et al., 2016).

Since the smartphone's functionality is the first priority, a few practices were developed to save resources (Kaufmann, 2016). If there is limited connection, no battery chargers available, or not enough bandwidth, using the smartphone is difficult (Alencar et al., 2018). The phone is only used for important tasks, if it is not needed, it is turned off at all times. Further, if groups of people are present, only one or a few phones are used to save battery life (Dekker et al. 2018; Kaufmann, 2016).

> We wanted to walk by ourselves instead of paying a smuggler. We used GPS to find the route. We started in Greece with a group of 75 persons. Five of them used GPS. We also contacted the first group [the group that left before them, HV] to find the way to Macedonia. So, I did not need a lot of information, except for the route and the others took care of that. I did not use internet myself. I only used my smartphone to take pictures.
>
> (Dekker et al., 2018, p. 6)

> When you find a place to charge the phone, you see 50 persons around it. . . . In Greece, we slept a night next to the Macedonian borders. There was a man who had a car with his wife; he had an engine from which there was a wire, so we gathered around it. You'd say a spider's web. I stayed 2 days without a phone because of battery. It was dangerous.
>
> (Gillespie et al., 2018, p. 5)

Consequently, smartphones are not used for entertainment purposes (Kaufmann, 2016; Emmer et al., 2020). Only if power supplies were nearby such as in hotels or busses, the phone is used to read or listen to music (Kaufmann, 2016). However, for fear of being detected, sometimes there were inhibitions to charge phones. Additionally, fears of phones being stolen or damaged (Zijlstra & van Liempt, 2017) or not being usable because of technical problems are prevalent (Dekker et al., 2018; Gillespie et al., 2016). Smartphones are also seen as an investment during migration (Ullrich, 2017; Gillespie et al., 2018). They are bought and sold, exchanged, fought over, or even gifted. If needed, smartphones can be exchanged for money or even used as a pledge for different parts of the journey (Ullrich, 2017). Not being able to use the phone even for a short amount of time can have devastating consequences: Not being able to deliver money to a smuggler on time, getting lost, and being separated from travel companions (Gillespie et al., 2018).

Akin to sharing a smartphone in groups, SIM cards are treated in a similar way. In every country, a new SIM card needs to be bought, making it rather expensive. Therefore, the hotspot function of the smartphone is used and everyone can use it (Kaufmann, 2016).

> After we went out from Greece it started to become difficult to use your phone because every country you have to buy a new sim-card . . . . It is difficult because we don't have papers and don't have passports so I think we were out from the internet in Eastern Europe until we came to Denmark but every time we had a chance to use the internet like maybe at the hotels or something then we use it. (Gough & Gough, 2019, p. 9)

As one asylum seeker puts it simply: "13 SIM cards. That's how I got to Europe" (Gillespie et al., 2018, p. 5).

According to Merisalo and Jauhiainen (2019), 85% of their study participants needed the Internet during their journey. Additionally, it is mainly used for interpersonal communication (Emmer et al., 2020). The Internet is not operated to actively search for information during the journey (as in accessing websites other than social media, for example) since information is disseminated through contacts (Emmer et al., 2016; Emmer et al., 2020). However, Frouws et al. (2016) describe how the Internet is used for every step of the way during migration, such as finding transportation or money transfer services. As the journey becomes more stable, for example, by reaching a crossover country such as Turkey or Lebanon, the Internet is used more often (Merisalo & Jauhiainen, 2019).

Concerning different applications and social media, a few insights could be made. The most important apps and websites considered are GPS (with offline possibilities such as Google Maps or Maps.Me), messaging services such as WhatsApp and Viber, and social media, especially Facebook and Facebook groups, as well as translation services such as Google Translate (Dekker et al., 2018; Emmer et al., 2020; Kaufmann, 2016). Even though non-government organizations (NGOs) developed websites and apps for refugees, none of the study's respondents knew about such offerings (Dekker et al., 2018).

One of the advantages of GPS is the application without Internet (Kaufmann, 2016). The best places to depart from and at what time can be determined (Zijlstra & van Liempt, 2017). Additionally, for accidents on sea, GPS coordinates can be shared with coast guards (Kaufmann, 2016). Of course, GPS can also be used to control routes suggested by smugglers and for any inconsistencies or attempted kidnappings (Gillespie et al., 2016). For example, the position was even checked around six to ten times a day to make sure everything was correct during the journey (Kaufmann, 2016). In relation to GPS, messaging services such as WhatsApp have additional benefits. With the option to send current positions of a person to another phone, missing or lost family members or travel companions can be found:

> In Hungary I bought a SIM card because I lost my friends. They took another bus so we were not in the same bus. My relatives and the sister of my wife and their spouses. Because of that I bought a SIM card to contact them. We found them in Budapest in the restaurant 'Istanbul'. . . . They sent me their position on WhatsApp so that we could find each other. (Kaufmann, 2016, p. 332, translated)

Similar to ICT and social media use before forced migration, asylum seekers apply similar methods during migration. Of course, messaging services are used to stay in contact with family and friends (Borkert et al., 2018; Kaufmann, 2016; Gillespie et al., 2016) and exchange information (Emmer et al., 2016) or to contact smugglers if needed (Gough & Gough, 2019). Accordingly, the best escape routes were identified by using social media such as WhatsApp, Facebook, and Viber and following trusted social ties who already migrated to Europe. Additionally, services such as WhatsApp, Viber (Gough & Gough, 2019), and Telegram were not subject to surveillance (Emmer et al., 2016). On social media like Facebook, groups help to exchange information and act as a swarm intelligence for gathering everything that could be relevant to reach a new destination in Europe (Kaufmann, 2016). There are even Facebook groups specifically created to travel without smugglers (Frouws et al., 2016). Further, these services and social media help to save money. "For example, when you use the internet, it would cost you 1500 euros to come here. When you don't use internet it would cost you around 5000 euros because you would need to pay smugglers et cetera" (Dekker et al., 2018, p. 6). Zijlstra and van Liempt (2017) describe a case where asylum seekers asked smugglers about different routes but declined his service in the end. They took the route themselves by relying on GPS and Google Maps. The line between asylum seekers and smugglers becomes blurry as asylum seekers are able to take others along: "Currently, it is so busy with refugees, and both the police and the refugees have more information about where to go. Now everyone is a professional. Now even the media on the television show the map of where you have to go" (Zijlstra & van Liempt, 2017, p. 184). However, the military sometimes censors details on these maps to protect borders, making it more difficult to find an efficient travel route (Gough & Gough, 2019).

Language skills are seen as an advantage during forced migration. However, due to the application of translation apps such as Google Translate, communication is made easier. For example, Zijlstra and van Liempt (2017) describe a case where two asylum seekers neither spoke English nor, in this case, Turkish, and stayed in a guesthouse in Turkey. When one of the asylum seekers downloaded the translation app, information could be exchanged between the guesthouse owner if necessary. The translation apps were also used at the borders and to ask for help if family members are ill or need medical care (Gillespie et al., 2016).

Similar to the findings about information needed before migration, this is mirrored by the information needed and collected during migration as well. As there are new circumstances that need to be adapted to, routes and methods are not always defined from the beginning and might change during the journey (Zijlstra & van Liempt, 2017). To find a safe travel route, this often means taking an already traveled route by other asylum seekers (Kaufmann, 2016). As others document their journey through social media and videos, they can serve as guidelines and help to react flexibly to new challenges (Ullrich, 2017). For this, asylum seekers believe in the suggestions of trusted relationships who already completed the journey. Information that is seen as important and spread among contacts along the routes are information related to travel routes, and how to satisfy basic needs such as food, security, or housing, prices of taxis, recommendable hotels, where to charge phones, where to buy equipment to sleep outdoors, which traffickers to trust, and how to avoid the police, as well as real time information about border situations (Ullrich, 2017). They connect deliberately and strategically with others that are already in another country or city, or at the border, to avoid dangers on the route by exchanging relevant information (Dekker et al., 2018; Frouws et al., 2016; Kaufmann, 2016; Zijlstra & van Liempt, 2017) and to counteract misinformation which can result in problems for asylum seekers (Alencar et al., 2018) or even death (Borkert et al., 2018; Gillespie et al., 2016). "We would keep contact through Viber with a group who left two hours before us. They would give us very up to date information. When you start your journey, you will receive the phone numbers of a hundred different people or so and when you have arrived, you don't know who they are any more" (Dekker et al., 2018, p. 6).

> In Aksaray, a lot of information is circulating: everyone tells each other things. On the third day that I was in Istanbul, an Iranian guy approached me and said 'I can take you to Europe for a small amount of money'. But I did not trust that, because why would this guy do such a risky thing for such a small amount? Smuggling has value, it should be somehow expensive. So there are a lot of people here that you cannot trust. We don't know anyone in Turkey, but with more people you have a stronger position, you can reach out and help each other if something happens. Therefore, it is better to stick together, as a group, and form a unity, especially when you come from the same town.
>
> (Zijlstra & van Liempt, 2017, p. 187)

Personal contacts and relationships are trusted the most (Kaufmann, 2016), such as other asylum seekers, volunteers, activists, or NGOs (Gillespie et al., 2018):

> On my journey there were two friends before me. They started the journey before me and when they arrived they called me. I talked with them: 'How did you do this? Is there police on the road? Maybe thieves?' They said: 'No, go to this smuggler.' They suggested: 'Don't go to this smuggler, but to this smuggler, he is better.'
>
> (Kaufmann, 2016, p. 331, translated)

During migration, group leaders are selected, which is most often connected to language and digital competencies such as being able to use smartphones effectively (Gillespie et al., 2018). Sometimes, these are even younger asylum seekers (Gough & Gough, 2019) as the older ones rely on them to be able to handle technology accurately (Dekker et al., 2018). Young or male asylum seekers are more likely to use the Internet than the older or female asylum seekers (Merisalo & Jauhiainen, 2019). This also means being able to change digital practices spontaneously – different smartphone applications need to be used to communicate and navigate or selectively deleted due to, e.g., low batteries. Other group members depend on them to lead the way on foot or even when crossing the sea by using, for example, navigation apps (e.g., Google Maps or Maps.Me):

> We were in the rubber boat, all the phones were in those little plastic bags we all buy, he [X] was the only one not to put his phone in a bag so he could stay in touch with coastguards and send our location to his brother in the Netherlands. Every few minutes, he used to tell his brother where we were. His brother was able to help guide us from a distance as he has already made the journey. (Gillespie et al., 2018, p. 7)

3.3 ICT and media practices in refugee camps

As mobile phone access was not considered a life sustaining need in 2015, costs of these kinds of services were not covered by organizers of refugee camps (Maitland & Xu, 2015). Because of this, earning money to pay for these services as well as other mobile technologies was necessary. This notion is supported by the high number of mobile phone (89%) and SIM card (85%) ownership. Other studies arrive at similar results, 279 out of 338 (83%) households in refugee camps use the Internet and almost all households have had at least one mobile phone (Fisher et al., 2019; Marume et al., 2018). Not surprisingly, due to the high coverage of mobile phones, this is the most common Internet access mode (Xu & Maitland, 2016). Xu and Maitland (2016) describe how mobile voice is used the most by the people residing in camps in contrast to Internet-enabled applications. If (mobile) Internet is not available due to poor network connectivity (Marume et al., 2018) or high costs (Merisalo & Jauhiainen, 2019), 2G is an alternative for communication (Hayes, 2019). Furthermore, the use of computers is decreasing. For example, Xu and Maitland (2016) describe a decrease from 24% in the homeland Syria to 10% in a refugee camp. A study conducted in 2018 in a refugee camp in Turkey arrived at another interesting insights (Smets, 2018). Here, shared technology was an important part of the media ecology. Whereas certain media are provided by the camp infrastructure, for example a computer room with desktop computers or television sets in common spaces, others were privately owned, but shared with family or other camp community members, which were cell phones or smartphones, laptop computers, or even television sets. Families would open their homes in the camp for others to watch TV, such as football or films (Smets, 2018). Similarly, the computer room was seen as a key-element in perceived well-being, not to write messages on computers, but because of social interactions as a lot of people gathered there (Smets, 2018). Furthermore, available technologies were used, besides for socialization and communication (Smets, 2018), for popular culture, such as online gaming (Yafi et al., 2018), music videos, football matches, soap operas, and films – as these may offer an escape from reality (Smets, 2018).

Owning a certain device would also lead to a particular status within the community – they can be exchanged for material goods or status. By owning a mobile phone, one can have access to the Internet – which consequently leads to the willingness to share time in the community and trying to help out (Fisher et al., 2019). Because, those with Internet access organize events and activities for the community. This can be attributed to the combination of education, language, and Internet access which suggests a form of self-efficacy fostered by those variables.

Social media is used to communicate with friends and family (Marume et al., 2018). Others apply social media such as Facebook for educational or economical purposes, for example, one participant offers Arabic language lessons through Facebook (Miconi, 2020). Further, news on political events can be exchanged through social media. Some even claim that social media is their main source for news: "[S]ometimes I got information faster through social media than my personal contacts" (Miconi, 2020, p. 6). This allows the users to debate about news: "[I]n social media there is freedom, freedom we don't have in Arab countries, political, of opinion, of speech, a possibility for expressing your own opinion" (Miconi, 2020, p. 5). There seems to be no consensus on the preferred social media platform for news consumption. Some participants claim: "Facebook is better than Twitter for gathering news" (Miconi, 2020, p. 5) and "people in Syria use Facebook to discuss political issues and their condition, and it is a more serious social site than Instagram" (Miconi, 2020, pp. 5–6). Others seem to prefer qualified sources found on Twitter, for example the official accounts of Al Jazeera or Al-Arabiya, allowing users to follow both sides of the situation in Syria. However, social media is also seen through a more critical point of view (Miconi, 2020). Here, personal contacts outside of social media are preferred:

> Usually my information about what happens [in] Syria don't come from social networks [SN], and I prefer normal phone calls in the normal mobile Turkish line, with acquaintances or friends. In less than 10% of cases my information come from there, because I think the information coming directly from the people you know are more reliable than what it is possible to find in SN. . . . I don't trust the information in social media, unless they are from sources you already know, you know they are reliable, for instance experts, or people they are known for their capacities. I prefer sources with an external credibility, beyond SN, rather in social media they are many people that take advantage of this for getting a popularity. (Miconi, 2020, p. 6)

A particular problematic point of view is also described:

> 'Even so, it seems that even correct information, in its turn, may become a problem: In SN clearly it is possible to observe what happens in your neighborhood and maybe sometimes it is better not to know, I saw that my home was destroyed during a bombing, like my grandparents' house.' What Tamara, Lilas, and Mushin Hussein reveal is the dark side of digital communication, where direct access to war information bypasses the gatekeeping function of traditional media, giving shape to an unstable environment.
> (Miconi, 2020, p. 6)

Conflicting results according the ownership of mobile technology as well as the application of digital media of people living in refugee camps can be observed. According to Maitland and Xu (2015), most youths (19 out of 21) do not have their own smartphone, 13 of these being younger than 18. However, a study conducted in 2015 and published in 2018 by Fisher and Yafi (2018) arrives at quite different results. Here, a survey showed that most youths do have, at least, access to ICT – 86% own mobile handsets and 83% own SIM cards. Further, social media is applied by them, with WhatsApp being the most used app and Google being the most used search engine. Further, youth are strong users of Facebook, YouTube, Skype, TV, and Wikipedia, and reported frequently helping with things related to ICT for family, friends, and others (Fisher & Yafi, 2018). According to Xu and Maitland (2016), male asylum seekers are more likely to use WhatsApp, Viber, Skype mobile voice, and e-mail than female asylum seekers. Mostly the young language literate women (in contrast to older women) apply a wider range of social media and communication applications. A study took a closer look at different age groups in the context of ICT mediated entertainment (Xu & Maitland, 2016). The younger participants are more interested in online entertainment activities than older ones, showcasing how young users are able to entertain themselves with ICT. Different social media services are applied for different needs (Miconi, 2020). Here, a clear difference could be made between people who used social media already in their homeland and those who discovered it in a new country or refugee camp. The first group encompasses skilled and curious youth who are able to use different platforms and services for various reasons and purposes. For example, youth keep personal pages on Facebook where they post stylized photos of themselves and friends at Za'atari refugee camp. They keep pages for local clubs and groups, resulting in a form of self-presentation. Page names reflect their strong feelings for Syria and hope for the future (Yafi et al., 2018). The second group is mainly interested in maintaining social ties, often using Facebook to do so. Many of the interviewed subjects discovered social media only after fleeing to a new country or refugee camp. Often, separation made it necessary to adopt social media (Miconi, 2020).

3.4 The special role of ICT and media

Overall, the smartphone is seen as one of the most important aspects during forced migration. Alencar et al. (2018) describe the smartphone as being a companion, organizational hub, a lifeline, and a diversion during forced migration. Gillespie et al. (2018) describe it as being central for mobility, being able to locate oneself and others, and safety. The smartphone prevents feeling disoriented, geographically, but surely emotionally as well (Kaufmann, 2016). By being able to contact friends and family, and even coast guards, a feeling of safety during stressful times can be fostered (Alencar et al., 2018). As one asylum seeker puts it simply: "Without my phone, I feel completely lost, stripped, naked, like missing a limb" (Gillespie et al., 2018, p. 6). In this context, someone else said that without a smartphone they would not know where to go or which direction to follow. To travel without a smartphone is like being in the desert – one just can't find the right path to reach the destination (Kaufmann, 2016, p. 338).

However, according to Dekker et al. (2018), social media information might not be as important for migration decision making as was previously assumed. Three reasons were mentioned. First, journeys were organized before migration: "I did not buy a SIM-card because my trip was already arranged before I left Syria. When I reached Greece, I went to a café and used the Wi-Fi that was available there" (Dekker et al., 2018, p. 6). Second, migration infrastructures developed along Europe's outer borders and major routes to Western Europe, making online information irrelevant: "The police helped us to take the bus and the train and the Red Cross also assisted us. I do not remember which countries we crossed.

I travelled together with others and we took the bus and then the train" (Dekker et al., 2018, p. 6).

Often, smugglers were met along the routes, offering to assist crossing the Mediterranean Sea. Third, the smartphone is used in groups, therefore not everyone needs to use a phone or the Internet. Emmer et al. (2016) argue that only half of those who used the Internet during their flight used applications such as Google Maps. All in all, social media is seen as helpful if the migration route was not planned before starting the journey and one needs to be flexible (Dekker et al., 2018). This also ties in with rumors heard during migration. These are more likely to influence decisions of those who do not have a fixed plan (Dekker et al., 2018).

It needs to be acknowledged that the country of origin influences ICT and smartphone usage and ownership because of economic factors. According to Emmer et al. (2016), around 80% of the Syrian and Iraqi asylum seekers owned a smartphone, whereas those from Central Asia, it was only around a third. Additionally, the Internet is used less by certain populations before and during the flight (Emmer et al., 2020). For example, asylum seekers fleeing from the Sub Saharan cannot afford new communication technologies. "The refugees come from different countries and two thirds from developing countries and countries where politics control the media. Communication technologies are not reachable for them, especially for those from the countryside. Poverty also plays a role. Electricity is luxury" (Ullrich, 2017, p. 6).

Concerns associated with social media and ICT usage need to be acknowledged as well. Fear of surveillance by government organizations by, e.g., leaving digital traces, were mentioned (Dekker et al., 2018; Gillespie et al., 2016; Wall et al., 2015). This is also related to the fear of being surveilled by extremist groups (Gillespie et al., 2016). When this kind of surveillance was observed, social media usage was stopped immediately. If coast guards or police were suspected to be nearby who could trace phone signals, the phone was turned off (Dekker et al., 2018). Further, smugglers sometimes forbid the use of smartphones and the Internet:

> We travelled with 13 people in the back of a truck. I did not know the route and we could not see anything. We were only allowed to leave the truck at night. We were not allowed to use our smartphones or any internet. The smuggler took our phones from us. He said that it would be safer for us, that no one would call the police for example.
>
> (Dekker et al., 2018, p. 7)

Also, smugglers are able to use social media to spread misinformation about safe routes. However, social media can also help to share information about these scams (Frouws et al., 2016).

Even though smartphones are important for information, just owning a mobile phone does not mean one can obtain relevant and important information, a term called “information precarity” (Wall et al., 2015, p. 1). Asylum seekers need to rely on underground sources since, it is argued, governments and news media do not provide timely and adequate information. Allegedly, the governments shied away from distributing accurate news during the height of migration since it was seen as a politically sensitive topic and fears of facilitating asylum seekers to come to Europe was prevalent (Gillespie et al., 2016). Additionally, although NGOs developed specific websites and apps for asylum seekers, these are not widely known or used (Dekker et al., 2018). This means there is either no awareness about these sites or no trust in the provided information. Frouws et al. (2016) however explain that aid organizations are aware of the reliance on ICT and social media by asylum seekers and actively use these to provide information on dangers and offer help on the routes.

All in all, factors such as smartphones used in groups and relying on opinion leaders who actively use applications such as GPS functions in conjunction with social media information to lead the way should be acknowledged. Surely, not all asylum seekers need to use a smartphone, but the reliance on these ICTs and applications is apparent.

References

Alencar, A., Kondova, K., & Riddens, W. (2018). The smartphone as a lifeline: an exploration of refugees' use of mobile communication technologies during their flight. *Media, Culture & Society*, *41*(6), 828–844.

Borkert, M., Fisher, K. E., & Yafi, E. (2018). The best, the worst, and the hardest to find: How people, mobiles, and social media connect migrants in(to) Europe. *Social Media + Society*, *4*(1), 1–11.

Dekker, R., Engbersen, G., Klaver, J., & Vonk., H. (2018). Smart refugees: How Syrian asylum migrants use social media information in migration-decision making. *Social Media + Society*, *4*(1), 1–11.

Emmer, M., Kunst, M., & Richter, C. (2020). Information seeking and communication during forced migration: An empirical analysis of refugees' digital media use and its effects on their perceptions of Germany as their target country. *Global Media and Communication*, *1*(2), 167–186.

Emmer, M., Richter, C., & Kunst, M. (2016). *Flucht 2.0. Mediennutzung durch Flüchtlinge vor, während und nach der Flucht* [Escape 2.0. Media usage by refugees before, during and after their escape]. Freie Universität Berlin. https://www.polsoz.fu-berlin.de/kommwiss/arbeitsstellen/internationale_kommunikation/Media/Flucht-2_0.pdf

Fisher, K. E., & Yafi, E. (2018). Syrian youth in Za'atari refugee camp as ICT wayfarers: An exploratory study using LEGO and storytelling. In E. Zegura (Ed.), *COMPASS '18:*

Proceedings of the 1st ACM SIGCAS Conference on Computing and Sustainable Societies (Article 32). Association for Computing Machinery.

Fisher, K. E., Yafi, E., Maitland, C., & Xu, Y. (2019). Al Osool: Understanding information behavior for community development at Za'atari Syrian refugee camp. In H. Tellioglu (Ed.), *C&T '19: Proceedings of the 9th International Conference on Communities & Technologies – Transforming Communities* (pp. 273–282). Association for Computing Machinery.

Fleay, C., Cokley, J., Dodd, A., Briskman, L., & Schwartz, L. (2016). Missing the boat: Australia and asylum seeker deterrence messaging. *International Migration, 54*(4), 60–73.

Frouws, B., Phillips, M., Hassan, A., & Twigt, M. (2016). *Getting to Europe the 'WhatsApp' way: The use of ICT in contemporary mixed migration flows to Europe.* Regional Mixed Migration Secretariat. https://mixedmigration.org/wp-content/uploads/2018/05/015_getting-to-europe.pdf

Gillespie, M., Lawrence, A., Cheesman, M., Faith, B., Iliadou, E., Issa, A., Osseiran, S., & Skleparis, D. (2016). *Mapping refugee media journeys. Smartphones and social media networks.* The Open University / France Médias Monde. https://www.open.ac.uk/ccig/sites/www.open.ac.uk.ccig/files/Mapping%20Refugee%20Media%20Journeys%2016%20May%20FIN%20MG_0.pdf

Gillespie, M, Osseiran, S, & Cheesman, M. (2018). Syrian refugees and the digital passage to Europe: Smartphone infrastructures and affordances. *Social Media + Society, 4*(1), 1–12.

Gough, H. A., & Gough, K. V. (2019). Disrupted becomings: The role of smartphones in Syrian refugees' physical and existential journeys. *Geoforum, 105*, 89–98.

Hayes, J. (2019). Trajectories of belonging and enduring technology: 2G phones and Syrian refugees in the Kurdistan region of Iraq. *European Journal of Communication, 34*(6), 661–670.

Kaufmann, K. (2016). How do refugees use their smartphones on the journey to Europe? Results of a qualitative interview study with Syrian asylum seekers in Austria [Wie nutzen Flüchtlinge ihre Smartphones auf der Reise nach Europa? Ergebnisse einer qualitativen Interviewstudie mit syrischen Schutzsuchenden in Österreich]. *SWS-Rundschau, 56*(3), 319–342.

Koikkalainen, S., Kyle, D., & Nykänen, T. (2019). Imagination, hope and the migrant journey: Iraqi asylum seekers looking for a future in Europe. *International Migration, 58*(4), 54–68.

Maitland, C., & Xu, Y. (2015). A social informatics analysis of refugee mobile phone use: A case study of Za'atari Syrian refugee camp. In *TPRC 43: The 43rd Research Conference on Communication, Information and Internet Policy Paper* (pp. 1–10). Available at SSNR: https://ssrn.com/abstract=2588300

Marume, A., January, J., Maradzika, J., (2018). Social capital, health-seeking behavior and quality of life among refugees in Zimbabwe: a cross-sectional study. *International Journal of Migration, Health and Social Care, 14*(4), 377–386.

Merisalo, M., & Jauhiainen, J. S. (2019). Digital divides among asylum-related migrants: Comparing internet use and smartphone ownership. *Tijdschrift voor Economische en Sociale Geografie, 111*(5), 689–704.

Miconi, A. (2020). News from the levant: A qualitative research on the role of social media in Syrian diaspora. *Social Media + Society, 6*(1), 1–12.

Musarò, P. (2019). Aware migrants: The role of information campaigns in the management of migration. *European Journal of Communication, 34*(6), 629–640.

Ramadan, R. (2017). Questioning the role of Facebook in maintaining Syrian social capital during the Syrian crisis. *Heliyon, 3* (12), Article e00483.
Smets, K. (2018). The way Syrian refugees in Turkey use media: Understanding "connected refugees" through a non-media-centric and local approach. *Communications, 43*(1), 113–123.
Ullrich, M. (2017). Media use during escape–A contribution to refugees' collective agency. *spheres: Journal of Digital Cultures*, *4*, 1–11.
Wall, M., Campbell, M. O., & Janbek, D. (2015). Syrian refugees and information precarity. *New Media & Society*, *19*(2), 240–254.
Xu, Y., & Maitland, C. (2016). Communication behaviors when displaced: A case study of Za'atari syrian refugee camp. In K. Toyama (Ed.), *ICTD '16: Proceedings of the Eighth International Conference on Information and Communication Technologies and Development* (Article 58). Association for Computing Machinery.
Yafi, E., Yefimova, K., & Fisher, K. E. (2018). Young hackers: Hacking technology at Za'atari Syrian refugee camp. In R. Mandryk & M. Hancock (Eds.), *CHI EA '18: Extended Abstracts of the 2018 CHI Conference on Human Factors in Computing Systems* (Article CS21). Association for Computing Machinery.
Zijlstra, J., & van Liempt, I. (2017). Smart(phone) travelling: Understanding the use and impact of mobile technology on irregular migration journeys. *International Journal of Migration and Border Studies*, *3*, 174–191.

4 Integration initiatives

While taking a closer look at different initiatives to help asylum seekers with the integration processes, first, the term integration needs to be defined:

> The integration of refugees is a dynamic and multifaceted two-way process which requires efforts by all parties concerned, including a preparedness on the part of refugees to adapt to the host society without having to forego their own cultural identity, and a corresponding readiness on the part of host communities and public institutions to welcome refugees and meet the needs of a diverse population. The process of integration is complex and gradual, comprising distinct but inter-related legal, economic, social and cultural dimensions, all of which are important for refugees' ability to integrate successfully as fully included members of the host society. (UNHCR, 2014, p. 1)

The host society is also responsible to create conditions that foster integration, e.g., giving access to services and employment as well as the acceptance of asylum seekers by society in the host society (Ager & Strang, 2008). According to Oduntan and Ruthaven (2019), there seems to be an order of needs expressed by asylum seekers and refugees for integration. Therefore, they propose an information needs matrix (in authors' own words: similar to Maslow's hierarchy of needs which describes the five stages of basic human needs) with informal tips for asylum seekers and operational points for the host society to help with the integration process. For example, education precedes information about social and employment information needs. Therefore, if the required language level and language information need is not met, the person may remain unemployed. According to Yu, Ouellet, and Warmington (2007), in 2007, several services were available to asylum seekers to help them orient themselves in a new environment. Here, three key areas related to these services were identified. First, the area of orientation, reception, and housing support: this aspect focuses on, for example, airport reception of asylum seekers, orientation to life in the new country, e.g., banking systems, culture, or providing temporary accommodation and support to find permanent accommodation; second, employment and language: job search techniques, English language conversation classes; and third, social capital, health and counseling, as well as family support including groups for specific purposes such as parenting classes, family counseling, women's groups, and furthermore, basic and emergency health cost coverage.

When orienting oneself in a new society, it was investigated how information literacy practices adapt in order to engage with a new, complex, and multimodal "information landscape" (Lloyd et al., 2013, p. 10). Mediators can share information with asylum seekers to help them become socially included. It is known that information literacy plays an important part in being able to participate in a

https://doi.org/10.1515/9783110672022-004

society. There are a lot of challenges faced by asylum seekers for which information is crucial: "Basic stuff . . . the housing, the employment, the income, the education . . . some won't understand money" (Lloyd et al., 2013, p. 16). Visual sources such as DVDs, magazines, "range of technologies" (Lloyd et al., 2013, p. 14), as well as social sources are seen as helpful when trying to adapt to a new life. It was stated that smartphones and digital or social media are seen as helpful for asylum seekers for integration (Emmer et al., 2016). In this scope, different ICT and media initiatives are presented.

Andrade and Doolin (2016) state that ICT can be a resource that offers five valuable potentials: participation in an information society, effective communication, understanding a different society, being socially connected, and expression of cultural identity. With those abilities, refugees can enhance their well-being and feel a sense of agency when establishing a new life. Digital technologies and ICT are also orientation devices with which migrants are able to imagine their lives elsewhere, beyond their home country (Twigt, 2018).

However, some governments do not seem to offer sufficient ICT services. A study concluded that ICT governmental services are not adequately developed, for example, there is a lack of official information sources in other languages. Furthermore, the websites are not culturally adapted. Online content should be made inclusive which especially targets the needs of different refugee and asylum applicant's populations (AbuJarour & Krasnova, 2017; Alencar & Tsagkroni, 2019). Benton and Glennie (2016) investigated apps, initiatives, and programs developed for integration. They found that governments should support the efforts of tech communities further so that relevant data can be accessed through apps more easily. Efforts to build applications are already made by scholars to help newly arriving asylum seekers to orientate themselves (Baranoff et al., 2015). It was also stressed that asylum seekers need faster and easier access to information about support systems or organizations. Information that is needed includes: access to electricity and Wi-Fi, access to healthcare, education and employment, and data security (Gillespie et al., 2018). Furthermore, ICT skills to use digital technologies efficiently are a growing requirement for various occupations in the majority of countries (OECD, 2016). Targeted courses for those skills could be beneficial for asylum seekers to make labor market integration easier (Stiller & Trkulja, 2018). ICT usage could also be beneficial for young asylum seekers. In an example of marginalized students, they were able to use technology to socialize with others and build friendships. This way, adolescents are able to express themselves (Gilhooly & Lee, 2014).

Another important aspect for asylum seekers is computer security. Since asylum seekers seem to have distinct computer security and privacy needs as well as constraints, technologies have to be designed for this community. This is especially difficult as there are cultural, language, and knowledge barriers that are in the way for recommended best practices (Simko et al., 2018).

In the following, some ICT-mediated solutions are presented. Three general aims can be identified based on an analysis of efforts made by ICT entrepreneurs to support integration. First aim: Helping newcomers navigate local services – several apps can help with the translation of information about local services. However, these apps do not always reach their target audience and often try to serve as an alternative to sometimes badly coordinated government services. Therefore, it is suggested to make government websites more accessible, multilingual, and responsive to user needs. Second aim: Getting newcomers into working or training – distance e-learning or employment-matching platforms are mentioned. Third aim: Providing access to community-based housing services – these services are aimed at matching natives of the new country with asylum seekers to exchange goods and services (Benton & Glennie, 2016). It is argued that smartphones are the only information access points many asylum seekers have. Therefore, they are used to apply geo-location services, learn a language, and contact friends and family. Consequently, it is requested that governments, industries, NGOs, and the local population should put more effort into providing adequate services utilizing such ICT products (AbuJarour & Krasnova, 2017).

Mason and Buchmann (2016) propose specific recommendations about ICT initiatives for asylum seekers: 1) Already existing ICT projects should be considered and expanded upon before starting a new one to save resources; 2) Barriers for asylum seekers need to be lowered, for example, using WhatsApp and Facebook to reach asylum seekers is far more effective than downloading a specific app; 3) Tech-access and literacy need to be increased as tech-literacy is not as pronounced among asylum seekers; 4) Focus on user-centered design; 5) Reaching asylum seekers is seen as the biggest hurdle; 6) Responsible data practices need to be prioritized; 7) Engaging the civic tech community is an important aspect; 8) Mutually complementary cooperation between different groups is seen as being helpful.

Some example projects and apps are described in the following paragraph. Concerning online education, understanding the preferences of asylum seekers when it comes to learning, e.g., a new language, face-to-face contact in context with learning is preferred over online learning. One main reason is that by being exclusively online, building social networks is difficult. Therefore, blended learning approaches are preferred (Castaño-Muñoz et al., 2018). As many asylum seekers already use different e-learning offers and learn languages through YouTube, Facebook, WhatsApp, and special apps, providing information this way could be beneficial. Furthermore, information through e-learning offers is typically presented in a simple format, making it more accessible (AbuJarour & Krasnova, 2018).

An app prototype called INTEGREAT was developed to help with everyday life challenges and to provide local information (Schreieck et al., 2017b). Information challenges identified by asylum seekers are a lack of information in general, the foreign culture, understanding guidelines, rural accommodations with bad infrastructures, and long waiting times at public institutions. The app provides information about language, work and education, family related issues, health, and everyday life. It is proposed that mobile-based solutions would be suitable to reach different groups as they can be easily translated to different languages. Furthermore, the next steps needed based on the number of days spent in the country can be highlighted. It should be made clear to asylum seekers what kind of information they are unaware of. Therefore, an initiation package was developed as a smartphone app containing all information and shortcuts needed for the first steps for integration (Irani et al., 2018). Another App called HELP@APP focuses on self-help for traumatized Syrian refugees arriving in Germany. This app aims at supporting those who do not have access to psychological care due to legal regulation, language barriers, and unclear cost coverage (Golchert et al., 2019). The app Moin, developed in Germany, focusses on informal learning through face-to-face communication and additional language learning features in addition with gamification elements to help teenagers with integration. By learning the language, social relationships can be built and integration made easier (Ngan et al., 2016). An e-mental health intervention program designed as an app, Step-by-Step (SbS), was proposed. To this end, information about user needs and feedback on a prototype was collected. As not all asylum seekers possess digital literacy skills, it was considered how to make such apps more accessible. Designing them similarly to already familiar applications, navigation can be made easier. For example, due to the great popularity of WhatsApp, the prototype resembled the user interface of this app (Burchert et al., 2019).

Similar efforts were made by providing a digital communication tool to help with informal learning and integration by providing personal dialogues between natives of the host society and asylum seekers. It is argued that it would be effective to create an online platform to support language learning and enable digital-mentorship in a social network (Hansson et al., 2017). Another digital information platform, INTEGREAT (the same as the above mentioned app), was developed in the context of nonprofit platform ecosystems to help bridge information gaps (Schreieck et al., 2017a). An online platform provided by the association *Refugees Welcome* is intended to connect asylum seekers with people offering temporary accommodations (Ferrari et al., 2020). Even digital collaborative games for entertainment are seen as beneficial in culturally diverse classrooms as they foster intercultural interaction and, in the long-term, integration (Alencar & de la Hera Conde-Pumpido, 2018).

References

AbuJarour, S., & Krasnova, H. (2017). Understanding the role of ICTs in promoting social inclusion: The case of Syrian refugees in Germany. In I. Ramos, V. Tuunainen, & H. Krcmar (Eds.), *Proceedings of the 25th European Conference on Information Systems (ECIS 2017)* (pp. 1792–1806). Association for Information Systems.

AbuJarour, S., & Krasnova, H. (2018). E-learning as a means of social inclusion: The case of Syrian refugees in Germany. In *Proceedings of the 24th Americas Conference on Information Systems (AMCIS 2018)* (pp. 2216–2225). Association for Information Systems.

Ager, A., & Strang, A. (2008). Understanding integration: A conceptual framework. *Journal of Refugee Studies, 21*, 166–191.

Alencar, A., & de la Hera Conde-Pumpido, T. (2018). Gaming in multicultural classrooms: The potential of collaborative digital games to foster intercultural interaction. In K. Lakkaraju, G. Sukthankar, & R.T. Wigand (Eds.), *Social Interactions in Virtual Worlds* (pp. 288–310).

Alencar, A., & Tsagkroni, V. (2019). Prospects of refugee integration in the Netherlands: Social capital, information practices and digital media. *Media and Communication*, *7*(2), 184–194.

Andrade, A. D., & Doolin, B. (2016). Information and communication technology and the social inclusion of refugees. *Management Information Systems Quarterly*, *40*(2), 405–416.

Baranoff, J., Israel Gonzales, R., Liu, J., Yang, H., &, Zheng, J. (2015). Lantern: Empowering refugees through community-generated guidance using near field communication. In B. Begole & J. Kim (Eds.), *CHI EA '15: Proceedings of the 33rd Annual ACM Conference Extended Abstracts on Human Factors in Computing Systems* (pp. 7–12). Association for Computing Machinery.

Benton, M., & Glennie, A. (2016). *Digital humanitarianism: How tech entrepreneurs are supporting refugee integration*. Migration Policy Institute. https://www.migrationpolicy.org/research/digital-humanitarianism-how-tech-entrepreneurs-are-supporting-refugee-integration

Burchert, S., Alkneme, M. S., Bird, M., Carswell, K., Cuijpers, P., Hansen, P., Heim, E., Harper Shehadeh, M., Sijbrandij, M., van't Hof, E., & Knaevelsrud, C. (2019). User-centered app adaptation of a low-intensity e-mental health intervention for Syrian refugees. *Frontiers in Psychiatry, 9*, Article 663.

Castaño-Muñoz, J., Colucci, E., & Smidt, H. (2018). From fragmentation to integration: Addressing the role of communication in refugee crises and (re)settlement processes. *International Review of Research in Open and Distance Learning*, *19*(2), 1–21.

Emmer, M., Richter, C. & Kunst, M. (2016). *Flucht 2.0. Mediennutzung durch Flüchtlinge vor, während und nach der Flucht* [Escape 2.0. Media usage by refugees before, during and after their escape]. Freie Universität Berlin. https://www.polsoz.fu-berlin.de/kommwiss/arbeitsstellen/internationale_kommunikation/Media/Flucht-2_0.pdf

Ferrari, M., Bernardi, M., Mura, G., & Diamantini, D. (2020). The potentials of digital collaborative platforms for the innovation of refugees' reception strategies. *Revista de Cercetare si Interventie Sociala*, *68*, 64–82.

Gillespie, M., Osseiran, S., & Cheesman, M. (2018). Syrian refugees and the digital passage to Europe: Smartphone infrastructures and affordances. *Social Media + Society*, *4*(1), 1–12.

Gilhooly, D., & Lee, E. (2014). The role of digital literacy practices on refugee resettlement. *Journal of Adolescent & Adult Literacy*, *57*(5), 387–396.

Golchert, J., Roehr, S., Berg, F., Grochtdreis, T., Hoffmann, R., Jung, F., Nagl, M., Plexnies, A., Renner, A., König, H.-H., Kersting, A., & Riedel-Heller, S. G. (2019). HELP@APP: Development and evaluation of a self-help app for traumatized Syrian refugees in Germany – a study protocol of a randomized controlled trial. *BMC Psychiatry*, *131*, 1–12.

Hansson, H., Qazi, H., Sundqvist, I., & Mozelius, P. (2017). Online digital mentorship: How might a digital communication tool facilitate informal learning and integration of newly arrived in Sweden. In A. Mesquita & P. Peres (Eds.), *Proceedings of the 16th European Conference on e-Learning* (pp. 178–184). Academic Conferences & Publishing International.

Irani, A., Bondal, P., Nelavelli, K., Kumar, N., & Hare, K. (2018). Refuge tech: An assets-based approach to refugee resettlement. In R. Mandryk & M. Hancock (Eds.), *CHI EA '18: Extended Abstracts of the 2018 CHI Conference on Human Factors in Computing Systems* (Article LBW554). Association for Computing Machinery.

Lloyd, A., Kennan, M. A., Thompson, K., & Qayyum, A. (2013). Connecting with new information landscapes: information literacy practices of refugees. *Journal of Documentation*, *69*(1), 121–144.

Mason, B., & Buchmann, D. (2016). *ICT 4 Refugees*. Deutsche Gesellschaft für Internationale Zusammenarbeit (GIZ) GmbH. https://regasus.de/online/datastore?epk=74D5roYc&file=image_8_en

Ngan, H. Y., Lifanova, A., Jarke, J., & Broer, J. (2016). Refugees welcome: Supporting informal language learning and integration with a gamified mobile application. In K. Verbert, M. Sharples, & T. Klobučar (Eds.), *Lecture Notes in Computer Science: Vol. 9891. Adaptive and Adaptable Learning* (pp. 521–524). Springer.

Oduntan O., & Ruthven I. (2019). The information needs matrix: A navigational guide for refugee integration. *Information Processing & Management*, *56*(3), 791–808.

OECD. (2016). *Skills for a digital world*. https://www.oecd.org/els/emp/Skills-for-a-Digital-World.pdf

Schreieck, M., Wiesche, M., & Krcmar, H. (2017a). Governing nonprofit platform ecosystems–an information platform for refugees. *Information Technology Enabled Collaboration for Development*, *23*(3), 618–643.

Schreieck, M., Zitzelsberger, J., Siepe, S., Wiesche, M., & Krcmar, H. (2017b). Supporting refugees in everyday life–intercultural design evaluation of an application for local information. In R. Alinda & P. S. Ling (Eds.), *PACIS'17: Proceedings of the 21st Pacific Asia Conference on Information Systems* (Article 149). Association for Information Systems.

Simko, L., Lerner, A., Ibtasam, S., Roesner, F., & Kohno, T. (2018). Computer security and privacy for refugees in the United States. In J. Li (Ed.), *2018 IEEE Symposium on Security and Privacy* (pp. 409–423). IEEE Computer Security.

Stiller, J., & Trkulja, V. (2018). *Assessing digital skills of refugee migrants during job orientation in Germany*. In G. Chowdhurry, J. McLeod, V. Gillet, & P. Willet (Eds.), *Lecture notes in computer science: Vol. 10766. Transforming Digital Worlds* (pp. 527–536). Springer.

Twigt, M. A. (2018). The mediation of hope: Digital technologies and affective affordances within Iraqi refugee households in Jordan. *Social Media+ Society*, *4*(1), 1–14.

UNHCR. (2014). *The integration of refugees. A discussion paper*. https://www.unhcr.org/cy/wp-content/uploads/sites/41/2018/02/integration_discussion_paper_July_2014_EN.pdf

Yu, S., Ouellet, E., & Warmington, A. (2007). Refugee integration in Canada: A survey of empirical evidence and existing services. *Refuge: Canada's Journal on Refugees*, *24*(2), 17–34.

5 German asylum procedure

By being granted *refugee status*, a person will be protected by international law, especially in accordance with the 1951 Refugee Convention. To determine which category of protection a displaced person belongs to, the Refugee Status Determination (RSD) process, managed by the government of the country of asylum or the United Nations High Commissioner for Refugees (UNHCR), will be conducted. This process is based on international, regional, or national law (UNHCR, 2019). As there is no established method on how to implement the RSD (with the exception of the 1951 Refugee Convention) every country's judicial and administrative system as well as the size or characteristics of the influx of displaced persons dictate the procedure. Since 2013, the UNHCR has followed the Handbook and Guidelines on Procedures and Criteria for Determining Refugee Status (UNHCR, 2019). The UNHCR helps 50–60 countries per year with the RSD process.

"Asyl ist in Deutschland ein von der Verfassung geschütztes Recht [Asylum is a right that is protected by the Constitution in Germany]" (BAMF, 2021a, p. 5). For the Federal Office for Migration and Refugees, the examination of various asylum applications is one of the most important duties. This requires expertise as well as responsibility and is a demanding task, since it is a complex procedure, involving different circumstances of the applicants and stringent legal frameworks. According to German law, there are four forms of protection: (1) entitlement to asylum, (2) refugee protection, (3) subsidiary protection, and (4) ban on deportation. Trained decision-makers from the Federal Office decide on each individual case which of the four grounds for protection apply that enables an applicant to remain in Germany. In the following, the different German legal frameworks will be illustrated. First, the steps after arrival will be described. This is followed by an explanation of the application processing at the Federal Office. Then, the special case of unaccompanied minors will be highlighted. This is preceded by particularities of the asylum procedure.

5.1 Asylum seekers: Proof of arrival

As soon as migrants arrive in Germany, they have to report themselves to a state organization (BAMF, 2021a). This can be done at the border or in the country, either to a security authority (such as the police), an immigration authority, a reception facility, or at an arrival center or AnkER facility (translated as Center for arrival, decision, and return). The AnkER facilities serve as a first accommodation for asylum seekers. Every individual will be registered in Germany and personal

https://doi.org/10.1515/9783110672022-005

data collected according to § 16 of the German Asylum Act (*Asylgesetz*; § 16 AsylG). Every applicant is photographed and fingerprints of people aged over 14 are taken. This data will then be stored in the Central Register of Foreigners which is managed by the Federal Office. The registered data is compared with already existing entries in the Central Register of Foreigners as well as the Federal Criminal Police Office. This way, it can be examined if an initial application or a follow-up application, among other possibilities, has been made. Furthermore, it can be assessed whether another European state is responsible for carrying out the asylum procedure by using the European Dactyloscopy system, or EURODAC, which is the European Union fingerprint database of asylum seekers and irregular border-crossers (European Parliament, 2013a).

Then, the proof of arrival according to § 63a AsylG is issued at an arrival center or reception facility, which will be valid for up to six months. The proof of arrival is the first official document and grants permit to settle in Germany for a while. Furthermore, this document grants permission to state benefits, such as accommodation, medical treatment, and food. But, this document is not an official identity card (BAMF, 2021a).

Table 5.1: EASY quota of the Federal States for 2015, rounded percentages. Source: BAMF (2015).

Federal State	Asylum application		Quota
	absolute value	percentage	Königsteiner Schlüssel
North Rhine-Westphalia	66,758	15.11%	21.24%
Bavaria	67,639	15.31%	15.33%
Baden-Wuerttemberg	57,578	13.04%	12.98%
Lower Saxony	34,248	7.75%	9.36%
Hesse	27,239	6.16%	7.32%
Berlin	33,281	7.53%	5.05%
Saxony	27,180	6.15%	5.10%
Rhineland Palatinate	17,625	3.99%	4.84%
Schleswig Holstein	15,572	3.52%	3.39%
Brandenburg	18,661	4.22%	3.08%
Saxony-Anhalt	16,410	3.71%	2.86%

Table 5.1 (continued)

Federal State	Asylum application		Quota
	absolute value	percentage	Königsteiner Schlüssel
Thuringia	13,455	3.05%	2.75%
Hamburg	12,437	2.81%	2.53%
Mecklenburg Western Pomerania	18,851	4.27%	2.04%
Saarland	10,089	2.28%	1.22%
Bremen	4,689	1.06%	0.94%

The asylum seekers are introduced to a nearby reception facility of the Federal State. It can be used temporarily or for long-term solutions up to six months, or until the application is decided on, depending on the circumstances. In some instances, the facility can be changed, for example to rejoin families. To calculate how many asylum seekers are distributed among the different Federal States, according to § 45 of the German asylum act (§ 45 AsylG), the EASY quota (translated as Initial Distribution of Asylum Seekers) system has been established (BAMF, 2021b). This quota is oriented on the *Königsteiner Schlüssel*, which takes tax income with 2/3 of the evaluation, and population with 1/3 of the evaluation as parameters for the calculation. This measure is adjusted annually to ensure a fair distribution among the 16 Federal States. As an example, for 2015, the highest quota was registered in the Federal State North Rhine-Westphalia (21.24%), followed by Bavaria (15.33%), and Baden-Württemberg (12.98%) (Table 5.1).

Overall, the facility is responsible for providing benefits or money for everyday personal needs. The amount of aid is regulated by the Asylum Seekers' Benefits Act (*Asylbewerberleistungsgesetz*; AsylbLG). The law was changed in 2015 because of rising numbers of asylum seekers in Germany. For example, instead of money, the benefits are given as payment in kind. Covered by this are food, housing, heating, clothing, healthcare and personal hygiene, household durables and consumables, other monetary assets to cover personal daily requirements, assistance in case of sickness, pregnancy and birth as well as further individual things, depending on the circumstances.

5.2 Asylum seekers: Personal asylum application

The personal asylum application needs to be filed at a branch office of the Federal Office, which is either an arrival center or an AnkER facility (BAMF, 2021a). To ensure sufficient communication, an interpreter will always be present when filling out the application. This also ensures that the applicants know their rights and duties regarding the procedure. Following official documents, all information is handed to them in their native language. Some form of identification needs to be provided if possible, which is either a national passport, or other personal documents such as birth certificates and driving licenses. The document will be screened by the Federal Office by physical and technical examination methods. Every application needs to be made in person. A written asylum application is only probable if the person in question is staying in a hospital or not of age.

5.3 Asylum applicants: Permission to reside

After the application was filed, and is still in processing, the applicant receives a certificate of permission to reside. With this document, asylum applicants can prove they are lawfully in Germany. But, the document is not the same as an official residence permit according to the German Residence Act (*Aufenthaltsgesetz*, AufenthG). Asylum applicants can reside in the same district in which the reception facility is located and are legally obliged to remain there (§ 56 para. 1 AsylG). To leave the area temporarily, permission is needed. This changes after three months and is expanded to cover the entire country (§ 63 para. 2 AsylG). If an applicant has poor prospects to stay in Germany, they must live in the reception facility until further decisions could be made. Assuming that the application is turned down as manifestly unfounded or inadmissible, the applicants are not allowed to reside anywhere else and are only allowed to temporarily leave the designated area if they got permission from the Federal Office until they leave the country. All asylum applicants who received the permission to reside are not allowed to partake in any kind of work for the first three months (§ 61 para. 2 AsylG).

Before the actual examination of the asylum application, the Dublin procedure is used to verify which EU Member State is responsible for the applicant (European Parliament, 2013b). For this, the Dublin III Regulation determines the criteria and procedures that need to be applied to find the responsible Member State for granting international protection. The first Member State where the fingerprints are taken or an application was made is responsible for the applicant (European Parliament, 2013b). The regulation is administered in all 28 EU Member States as well as in Norway, Iceland, Liechtenstein, and Switzerland. The Dublin procedure begins at the Dublin Center of the Federal Office if indications can be found that another Member State could be responsible for an application. If those indications are valid, a transfer request is addressed to the concerned Member State. In case that the Member State approves, the applicant will be sent to the Member State and the application therefore becomes inadmissible in Germany. The applicant can vote against this decision by applying to an administrative court. The immigration authorities and Federal Police are responsible for setting the transfer date and the relocation itself.

5.4 Asylum applicants: Personal interview

Then, the most important step in the procedure follows: the personal interview. Organizations that give advice on how to prepare for this part of the procedure sufficiently are available. Every person needs to be interviewed in person by a decision-maker of the Federal Office (§ 24 para. 1 AsylG). For special cases such as abused, tortured, or traumatized applicants, specially trained decision-makers will conduct the interview. To bridge any potential language barriers, interpreters are always present. If an applicant is not attending the interview without prior notice, the asylum procedure can be ended and the application be rejected. During the interview, a lawyer or representative of the UNHCR, or for unaccompanied minors a legal guardian, can be present to assist. Aim of the personal interview is to gather information on why the person fled their home country and to detect possible contradictions (§ 25 para. 1 AsylG). Asked about are the persons' curriculum vitae, living conditions, residencies, travel routes, stays in other countries, and if the person already applied for asylum in Germany or another country. If available, the applicant can provide proof, such as pictures, documents from police or other offices, or medical certificates. The person is also asked about the possible conditions of their home country that would await them if they return. Further reasons on why the applicant should not be deported need to be stated (§ 25 para. 2 AsylG). The interview is logged and translated so the applicant can make additions or revisions.

5.5 Decision on asylum status: Persons entitled to protection and persons entitled to remain

Based on the interview and a detailed examination of documents and items of evidence, the Federal Office decides on the asylum application. Since a lot of information is needed for this complex process, sometimes, further research needs to be conducted. Access to information is available via the information system of the Federal Office called MILo (translated as Migration Infologistics) (BAMF, 2014). This database stores information on the current migration and refugee situation in the world. Furthermore, individual requests to the Federal Foreign Office can be made, as well as language- and text analyses (for identity checks), physical-technical examinations of certificates, and the retrieval of medical or other reports are available.

Table 5.2: Forms of protections and legal basis. Source: BAMF (2021a).

Form of protection	Legal basis (English translation)
(1) Entitlement to asylum	§ 16a Grundgesetz (The Basic Law)
(2) Refugee protection	§ 3 AsylG (Asylum Act)
(3) Subsidiary protection	§ 4 AsylG (Asylum Act)
(4) Ban on deportation	§ 60 V + VII AufenthG (Residence Act)

Based on the German Asylum Act (AsylG), The Basic Law, and Residence Act (AufenthG), one of the four forms of protection will be granted (Table 5.2). This decision is forwarded to the applicant and the concerned immigration authorities. (1) Entitlement to asylum is granted to applicants who would otherwise be subject to serious human rights violations if they would return to their home country. The applicants are deemed to have been persecuted on political grounds because of race, nationality, political opinion, fundamental religious conviction, or membership of a particular social group, which can be, e.g., sexual orientation. Furthermore, they do not have an alternative of refuge within their country of origin. Exempt from this are people who came to Germany from a safe third country, which are the Member States of the European Union, Norway, and Switzerland. (2) Refugee protection is the extensive version of the entitlement to asylum and includes, additionally to the reasons mentioned, the persecution by non-state players. Persecution includes the use of physical or psychological violence; legal, administrative, judicial measures which are discriminatory; punishment; refusal to provide judicial legal protection; acts linked to sexuality or which target children. (3) Subsidiary protection is granted if neither refugee protection nor entitlement to asylum can be applied but if serious harm is threatened in the home country. This can be originating from governmental and non-governmental players. These threats are: the imposition of the death penalty; torture or inhuman punishment; a serious individual threat to the life or integrity of a civilian within international or armed conflict (§ 4 AsylG). Furthermore (4), a ban on deportation can be issued. This is the case if a return to the home country constitutes a breach of the

European Convention for the Protection of Human Rights and Fundamental Freedoms, or a considerable danger to life, limb, or liberty exists. An example for concrete danger could be related to health reasons, for example if the return could cause life-threatening or serious diseases to become much worse (§ 60 V + VII AufenthG).

5.6 Special case of unaccompanied minors

Every applicant under the age of eighteen is regarded as a minor in Germany. If they arrive without an adult being responsible for them or are left in Germany without a caretaker, they are regarded as an unaccompanied minor. The children are taken into care by a youth welfare office. Then, a suitable person, e.g., relatives or foster families as well as proper facilities specializing in caring for unaccompanied minors or youth welfare facilities will care for the child. If close social ties between unaccompanied minors exist, they can be accommodated together. Every minor will be appointed a guardian decided by the Family Court. This guardianship lasts as long as the minor reaches the age of maturity of his or her home country. The guardian or the youth welfare office needs to write the asylum application for the minor as they are not legally allowed to do so by themselves. As unaccompanied minors are a vulnerable group, they have special guarantees for their asylum procedure (BAMF, 2021a).

5.7 Rejections

The Federal Office has the right to reject asylum applications under different circumstances (§ 29 AsylG). When the Dublin Procedure was successfully concluded and Germany determined as the country responsible for the application, different scenarios can prohibit the asylum procedure. This may be the case if an applicant was already granted protection in another country, which could be, e.g., Member States of the EU, Norway, or Switzerland. If a relocation procedure took place, meaning that a person was sent from a refugee camp to another European country, the application that was filed in Germany is inadmissible. Furthermore, if an application was rejected, a new one cannot be made if there are no valid proofs or arguments that can be demonstrated to the Federal Office.

If none of the four forms of protections is assigned, the applicants receive a negative notice as well as a notice of intention to deport (§ 35 AsylG). There are two types of rejection: (1) outright rejection and (2) a rejection termed as "manifestly unfounded" (§ 30 AsylG). (1) if the outright rejection notice was received, the person has thirty days to leave Germany (§ 38 AsylG). If the person did not receive any protection and cannot be deported because of legal, humanitarian, or personal reasons, e.g., if the person in question is severely ill or does not have valid identification papers, the immigration authority can issue a temporary suspension of deportation (*Duldung*, § 60a para. 4 AufenthG), or even a residence permit. The Duldung status is not a form of protection and does not include the official permission to reside or stay in Germany. It is only a temporary suspension of deportation. This means that the applicant is technically obliged to leave Germany. After eighteen months, a residence permit should be granted but this depends on several circumstances: if the applicant is not able to leave the country because of reasons outside of his or her control or a voluntary departure would be impossible or infeasible, the permit will be authorized. The requirements according to the ban on deportation (§ 5 AufenthG) also need to apply. For individuals without an ID issued by their home country, the right of residence will not be granted (§ 60b AufenthG). (2) Reasons for a rejection termed "manifestly unfounded" could be: the applicant was not believed based on contradictions or fake evidence; a nationality was falsely stated or not stated; the application was filed a long time after arrival; economic circumstances are the reason for the application; the applicant is a danger to society, was sentenced to three years in jail, or is under suspicion to have committed a war crime, a crime against humanity, or a severe non-political crime. The applicants therefore have to leave the country as soon as possible. Every applicant who was rejected based on these reasons has the right to take court action against the decision within a week (§ 74 AsylG). The application will be examined by the court and a possible rejection or protection will be decided on.

Nonetheless, all forms of protection as well as rejection or Duldung status have different consequences for the person in question. A residence permit will be issued so that a person can live in Germany lawfully for up to five years. If an applicant wants to live in Germany for an indefinite period, a settlement permit is needed. This will only be granted under special circumstances (§ 9 AufenthG).

5.8 Family reunification

As the laws and regulations on family reunification are very complex (Informationsverbund Asyl und Migration e.V., 2018) the basics will be outlined in the following section. Depending on where the family members are currently living, different processes apply. There are two different forms:

(1) Visa procedure: This applies regardless where the family members live
(2) Dublin regulation procedure: Applicable if family members already arrived in a European country

There are different scenarios that play a role in this procedure. If it is the case that the applicant's family is still in the country of origin or another country outside Europe, only the husband or wife, or unmarried children under the age of eighteen are allowed to take part in the family reunification process. If the child applicant is underage, they are allowed to bring their parents. The other case applies if the family already arrived in a European country. Then, because of the Dublin Procedure, family members, in some cases including children over the age of eighteen, siblings, aunts, uncles, and grandparents, can be reunified. This applies if the person who wants to bring the family possesses a residency permit or is in the middle of the asylum procedure. To start the family reunification process, the European state in which the family members who want to come to Germany are located is responsible to do so. The family members have to submit an application for asylum to the state and inform the authority that they want to move to Germany.

The visa procedure can be applied under different circumstances and depends on the residency status of the person who wants to bring the family to Germany. Relevant criteria for this are long duration of separation of family members, separation of families with at least one (minor) unmarried child, serious risks to life, limb or personal freedom of a family member living abroad or a serious illness, or need for care or serious disabilities of a family member living abroad. Additionally, special focus is set on the welfare of children and further integration aspects, which include language skills or the ability to provide for a means of living (Informationsverbund Asyl und Migration e.V., 2018). If persons are concerned that pose a risk, incite hatred against groups of society, run a forbidden organization, or take part in violence in pursuit of political and religious grounds, they are excluded without exception. Persons with a national ban on deportation need to fit certain requirements and provide sufficient living spaces as well as a dependable income to bring their family to Germany. If one receives the Duldung status they are not allowed to transfer their family. For people with a permission to remain during the asylum procedure, only the Dublin Procedure can be applied, therefore the family members need to live in Europe.

5.9 Family asylum

Every person who is regarded as a family member and stays in Germany will be entitled to asylum, refugee protection, or subsidiary protection based on another family member's granted asylum, refugee protection, or subsidiary protection (§ 25 AsylG). Persons to whom a national ban on deportation has been administered do not have the right to family reunification. Defined as a family member are spouses and registered partners, minors, unmarried children, the parents of a minor, unmarried persons for the purpose of care and custody, other adults who have personal custody of a minor, unmarried persons, as well as minors, unmarried siblings of a minor (BAMF, 2021a). The application must be made with the Federal Office within three months after the person has been granted some form of protection.

Special cases are made for children who were born in Germany (§ 14a, 43 AsylG). If the parents filed for an asylum application before the baby was born, the child is given a separate asylum procedure. The application is automatically filed by the Federal Office and the parents are able to give individual reasons for the child's asylum application. If not, the same as the parents' reasons apply. If the application has been rejected, minor children will not be returned separately from their parents.

References

BAMF. (2014). Central access to important documents. MILo – the information system of the Federal Office. *The MILo database*. Retrieved from https://www.bamf.de/DE/Behoerde/Informationszentrum/MILo/milo-node.html;jsessionid=264E442E18B99EC206D7616D86231B22.internet531

BAMF. (2015). *Das Bundesamt in Zahlen 2015. Asyl, Migration und Integration* [The Federal Office in figures 2015. Asylum, migration and integration]. https://www.bamf.de/SharedDocs/Anlagen/DE/Statistik/BundesamtinZahlen/bundesamt-in-zahlen-2015.pdf?__blob=publicationFile&v=16

BAMF. (2021a). *Ablauf des deutschen Asylverfahrens. Ein Überblick über die einzelnen Verfahrensschritte und rechtlichen Grundlagen* [The stages of the German asylum procedure. An overview of the individual procedural steps and legal bases]. https://www.bamf.de/SharedDocs/Anlagen/DE/AsylFluechtlingsschutz/Asylverfahren/das-deutsche-asylverfahren.pdf?__blob=publicationFile&v=12

BAMF. (2021b, April 20). *Erstverteilung der Asylsuchenden (EASY)* [Initial distribution of asylum-seekers (EASY)]. BAMF. https://www.bamf.de/EN/Themen/AsylFluechtlingsschutz/AblaufAsylverfahrens/Erstverteilung/erstverteilung-node.html

European Parliament. (2013a). *Document 32013R0603*. European Parliament. https://eur-lex.europa.eu/legal-content/DE/TXT/?uri=CELEX:32013R0603

European Parliament. (2013b). *Document 32013R0604*. European Parliament. https://eur-lex.europa.eu/legal-content/en/TXT/?uri=CELEX:32013R0604#d1e1042-31-1

Informationsverbund Asyl und Migration e.V. (2018). *Advice on family reunification: March 2018*. https://familie.asyl.net/fileadmin/user_upload/pdf/MerkbFamZf_engl_201805_fin.pdf

UNHCR. (2019). *Handbook and guidelines on procedures and criteria for determining refugee status*. https://www.unhcr.org/publications/legal/5ddfcdc47/handbook-procedures-criteria-determining-refugee-status-under-1951-convention.html

6 Theoretical foundations

As the analysis of asylum seekers' information behavior is a very current and highly topical research area, there is no established theory to explain and to understand the interrelationships between asylum seekers (including the aspects of gender and generation), their information behavior, information needs, their level of information literacy, and their usage of information and communications technology (ICT) as well as of offline and online media. Due to prevailing circumstances, the asylum seekers were forced to leave their home country and thus had to move to a new country and environment – making it important to study their information horizons in the old and novel situation, while concentrating not only on information sources like social media, other online and offline media, and ICT, but also on offline and online information exchange with other people to satisfy newly occurring and targeting information needs. As theoretical foundations, mainly two approaches were applied, namely Information Behavior Research (with a research tradition especially in information science) and the Uses and Gratifications Theory (with a long research tradition in media and communications science). Additionally, it needs to be addressed what "media" or information channels are.

In line with the subtitle of this monograph, "New Life – New Information?", the primary purpose of this research is to investigate the information behavior as well as the social media, offline and online media, and ICT usage of asylum seekers and refugees in a new country. As the researchers are from Germany and there was a massive migration flow from countries of the Asian Middle East around 2015, this study comprises Syria, Iraq, Afghanistan, and Iran as asylees' former home countries and Germany as the target country, but – of course – for comparison also other parts of the world have been considered. The asylees' information behavior and applied social media, offline and online media, and ICT usage prior to forced migration, and therefore, in their home country and furthermore their practices in the target country are represented. Further, the differences between the homeland and Germany will be highlighted. For ICT and media usage, also gender-dependent and age-dependent differences and similarities for asylum seekers will be displayed. Other aspects investigated and studied are the information needs of asylum seekers and refugees as well as experiences of experts regarding the information behavior and ICT, online and offline media, and social media usage of asylum seekers.

The goal of this chapter is the development of a theoretical framework establishing a set of research questions as well as a research model. Anfara Jr. and Mertz (2015, p. 15) define theoretical frameworks "as any empirical or quasi-empirical theory of social and/or psychological processes, at a variety of levels

https://doi.org/10.1515/9783110672022-006

(e.g., grand, midrange, explanatory), that can be applied to the understanding of phenomena." Following, theoretical phenomena and concepts are combined as a theoretical framework and are further transformed into a research model. Furthermore, theoretical frameworks serve to present phenomena and interrelationships in an understandable and illustrative way.

For the theoretical framework of this research, firstly the concept of "information behavior" will be described. In this context, information needs, information production, information seeking and consumption as well as information horizons will be elaborated on. Furthermore, theories of media usage as the Uses and Gratifications Theory (Katz et al., 1973; Blumler & Katz, 1974) are consulted to describe the motives of media and ICT users for using certain media. Information (in the sense of knowledge) is one of the four fundamental aspects of the Uses and Gratifications Theory, along with entertainment, socialization, and self-actualization. Then, the used classification of media, which are used as information channels by asylees, will be introduced. The core part of the theoretical foundation is a model by Zimmer, Scheibe, and Stock (2018), which initially represents the information behavior of users on a synchronous social media service, i.e. social live streaming services. But the model "is (with small changes) suitable for all kinds of social media" (Zimmer et al., 2018, p. 444). This model is modified for this study's aim to represent and to understand the phenomenon of asylees' information behavior and their social media, online media, offline media, and ICT usage.

6.1 Information behavior research

Analyzing the information behavior and thereby the information needs as well as the information production, the information seeking and the information reception behavior as well as, additionally, the information horizons of people is an important aspect in information science studies. As the present research aims to investigate the information behavior as well as information needs of refugees and asylum seekers prior to escaping from their home country as well as after arriving in the new country and thereby integrating into and adapting to an unfamiliar society, the concepts of information behavior need to be elaborated on.

Information production is the process of creating and communicating parts of information by individual persons. With *information seeking*, such activities which contain the active search for information, e.g., by asking other people, to use a search engine or to consult a social media channel can be considered. *Information reception* may be active or passive. Active information reception is the application of actively sought information. In many situations, information is consumed passively (Bates, 2002; Wilson, 1997; Wilson, 2000). For example, the passive consumption of information happens during human interaction, resulting in information exchange (Wilson, 1981). Information behavior depends on context and situation and on one's personal information horizon (Sonnenwald, 1999; Sonnenwald & Iivonen, 1999).

In 1968, Paisley (1968) already mentioned how information science can meet behavioral science when studying information needs and information uses. The term *information behavior* is defined as an "umbrella concept" for describing the "ways that people generally deal with information" (Savolainen, 2007, p. 109). Wilson (2000) defines information behavior in a rather broad way with attention to information seeking and information use: "Information Behavior is the totality of human behavior in relation to sources and channels of information, including both active and passive information seeking, and information use. Thus, it includes face-to-face communication with others, as well as the passive reception of information as in, for example, watching TV advertisements, without any intention to act on the information given" (p. 49). According to this definition, all activities related to information are included in the concept of information behavior. People are searching for information although and while they do not actively perceive their acting as information seeking. Further support is provided by a statement from Bates (2002), she also mentioned that information seeking includes "all the information that comes to a human being during a lifetime, not just in those moments when a person actively seeks information" (p. 3). Spink and Cole (2004) define information seeking as a subset of information behavior that includes the purposive seeking of information in relation to reaching a goal and therefore satisfying a need. Although information seeking appears to be intentional, a person does not necessarily perceive an information need as active (Wilson, 2000). Case (2007) illustrates the context of information seeking and information needs in the following way: "Every day of our lives we engage in some activity that might be called information seeking, though we may not think of it that way at the time. From the moment of our birth we are prompted by our environment and our motivations to seek out information that will help us meet our needs" (p. 18). Stock and Stock (2013) write about the process of seeking information for all basic human needs: "However, many fundamental human needs require information in order to be satisfied. This begins with information seeking in order to locate food, water, and sex, and ends with problem-solving, which can only be achieved via knowledge (to be sought and retrieved)" (p. 469). The fundamental and basic human needs are coming from a person's psychological nature. Other needs people may recognize are safety needs, love and belonging needs, esteem needs and, finally, self-actualization (Maslow, 1954).

In 1981, Wilson (1981) investigated a model to display the interdependencies between the various concepts as well as different constituents of the information behavior area; the model was updated in 1999 (Wilson, 1999). According to the model, the information user is looking for some information because of a certain (information) need. As Stock and Stock (2013) also state, "[a]n individual's information need is the starting point of any search for information" (p. 469). The information seeking behavior results from using distinct techniques: Making demands on an information system (e.g., retrieval systems, online services, libraries) or other information sources as well as information exchange with other people. This process can result in success or failure for the relevance of the sought information. Afterwards, the information use may lead to information transfer, if exchanging information with other people, or the use of an information system, if successful, will satisfy or will not satisfy the information need of the user. As the model shows, information behavior is a collective term, including information need, information production, information seeking, information use, and further elements.

Hence, this approach does not only consider information seeking and consumption behavior (as often found in information science, e.g., Cole, 2012; Fisher & Julien, 2009) but information production and dissemination behavior as well. The definition by Pettigrew, Fidel, and Bruce (2001, p. 44) who describe "information behavior" as "how people need, seek, give, and use information in different contexts, including the workplace and everyday living" are used for this study.

Case and Given (2018) state, "context and situation are important concepts for information behavior research" (p. 48). Human information behavior is embedded in one's "*information horizon*" (Sonnenwald, 1999, p. 8) including social contacts and networks (their social capital) as well as their concrete contexts and situations. Sonnenwald (1999) defines *context* as "the quintessence of a set (or group) of past, present and future situations" (p. 3). Contexts are, for instance, academia, family life, citizenship, clubs, etc. Following Sonnenwald (1999), contexts have boundaries, constraints, and privileges which are perceived differently by participants and outsiders. A *situation* is characterized as "a set of related activities, or a set of related stories, that occur over time" (Sonnenwald, 1999, p. 3). *Social networks* "help construct situations and contexts, and are constructed by situations and contexts" (Sonnenwald, 1999, p. 4). They refer to communication among people, to patterns of connection, and to human interaction. Within the context and the situation there is an information horizon in which people act. When a person has decided to produce, to look for, or to receive information, there is an information horizon in which she or he realizes her or his information behavior. Sonnenwald (1999) only considers information seeking behavior, while we broaden this approach to all aspects of information behavior.

The success (or failure) of one's information behavior depends on her or his level of *information literacy* (Stock & Stock, 2013, chapt. A.5). According to UNESCO and the IFLA (2005), information literacy "empowers people in all walks of life to seek, evaluate, use and create information effectively to achieve their personal, social, occupational and educational goals." Information literacy includes two aspects. The first building block deals with practical competences for information retrieval starting with the recognition of an information need, proceeding via the search, retrieval and evaluation of information, and leading finally to the application of found and evaluated information (ACRL, 2000). The second building block summarizes practical competences for information production and knowledge representation. Apart from the creation of information (texts, images, audio files, and videos), it emphasizes their uploading, indexing, and storage in digital information services (e.g., on a social media channel), if necessary (Stock & Stock, 2013, p. 80). For both of these building blocks, it is of great use to possess basic knowledge of information law and information ethics.

Multiple aspects shape the information horizon of a user which results in human information behavior. As both, context and situation in which asylum applicants behave changed due to the escape of their home country and the resettling into the new country with a foreign language and a different culture, it is necessary to study their information literacy, the information needs, and the information horizons. Furthermore, how the asylum applicants satisfy their needs and therefore seek information is another aspect that should be considered in this research. On social media and messaging services both, information production as well as information seeking and information reception behavior, is always given. Besides social media, face-to-face communication as well as ICT devices and other media serve as information exchange sources and are considered in this study as well. But why are people using certain kinds of media? What are their motives?

6.2 Uses and gratifications theory

One of the classical sender-centered models in communication and media science is the theory of Lasswell (1948, p. 37), introducing the following questions: Who? Says What? In Which Channel? To Whom? With What Effect? Braddock (1958) adds two further questions: What circumstances? and What purpose? The extended Lasswell formula reads as follows: "WHO says WHAT to WHOM under WHAT CIRCUMSTANCES through WHAT MEDIUM for WHAT PURPOSE with WHAT EFFECT" (Braddock, 1958, p. 88). In terms of Braddock, the *Who* is the communicator, the *What* is the message with the two inseparable aspects of content and presentation, *To Whom* asks for the audience and its characteristics, *What Circumstances* analyzes the environment of the information behavior in terms of time and setting, *What Medium* includes questions on the information channel (online or offline), *What Purpose* means the communicator's motives to communicate, and, finally, *What Effect* analyzes the outcomes of the entire communication process for the audience. Interestingly, Braddock only makes mention of the motives of the communicator, but not of the audience.

Here, another theoretical framework, now audience-oriented, enters the stage. The Uses and Gratifications approach had its beginnings in 1962 when Katz and Foulkes (1962) started to ask what people do with the media instead of the question what media does with the people as Lasswell and Braddock asked. To understand why people use media and what they use the media for, Katz, Blumler, and Gurevitch (1973) summarized findings and described fundamental objectives of uses and gratifications research in media consumption, which resulted in the Uses and Gratifications Theory (U>). It is a popular method in mass communication studies to explain motives for media use and to identify what needs are satisfied. There are five fundamental assumptions for the uses and gratifications model (Katz et al., 1973):

1. The audience engages actively while dealing with media; the usage is goal-orientated as media consumption is shaped by expectations towards the media's content.
2. The recipient uses the media to satisfy his or her needs and decides actively whether to apply a specific medium or not.
3. There is a competition between the media and other sources, which are not media-based. Those alternatives may lead to need satisfaction as well.
4. Goals of media consumption can be derived by simply asking the users or confronting them, as they are aware of their motives.
5. Suspension of judgements about the value of the cultural significance of media while discovering audience orientations solely.

During the process of studying uses and gratifications, researchers "are concerned with: (1) the social and psychological origins of (2) needs, which generate (3) expectations of (4) the mass media or other sources, which lead to (5) differential patterns of media exposure (or engagement in other activities), resulting in (6) need gratifications and (7) other consequences, perhaps mostly unintended ones" (Katz et al., 1973, p. 510). Research has shown that three sources for need gratification and need satisfaction are given – those are media content, media exposure, and the social context as well as the situation of the media exposure (Katz et al., 1973).

McQuail, Blumler, and Brown (1972) identified four basic needs that are satisfied through media consumption – diversion (escaping reality, responsibilities, and problems as well as for relaxation); personal relationships (social contacts and substitute companionship); personal identity (self-realization);

and surveillance. Furthermore, 35 more detailed needs have been defined and identified by Katz, Gurevitch, and Haas (1973). Later, McQuail (1983) summarizes four central motives for media use of the 35 needs, which are: information, entertainment, social interaction, and self-identity.

For user-generated content, mostly on social media, Shao (2009) differentiates between three different user roles, which are consumers, participants, and producers. Users of social media, who are consuming, use it mostly for information and entertainment purposes, whereby participating users, those who are, for example, writing comments, have the goal to interact with others and to get integrated into the community. Finally, the users who are actively producing their own content and sharing it online, are using the media mainly for self-expression and self-actualization.

Palmgreen, Wenner, and Rayburn (1980) address the gap between "gratifications sought" and "gratifications obtained" in uses and gratifications research, as most studies are not differentiating between these two concepts. Earlier, in 1974, Greenberg (1974) described the phenomenon as "gratifications sought" and "gratifications received." It is the distinction between expectations of the audience on the media and its content (gratifications sought) and the need for satisfaction achieved by media consumption (gratifications obtained or gratification received). Also, there is a feedback loop, the media user may seek information, but will receive entertainment from the media (Palmgreen et al., 1980).

Latest research about uses and gratifications in social media usage claim that uses and gratifications have been shifted and changed for "new" (i.e., social) media types (e.g., Sundar & Limperos, 2013). In order to be able to compare the motives of using certain ICT or online, and offline as well as social or traditional media, the four central motives (information, entertainment, social interaction, and self-presentation) are applied in this study. Thereby, it was distinguished between socializing via communication channels and socializing in order to be able to speak about the received information. Furthermore, it was also differentiated between self-presentation by producing content or playing digital (online) games and personal identity, meaning to work on and build one's own personal identity, resulting in information (knowledge), entertainment, socializing or socializing: communicating, and self-presentation or personal identity.

6.3 ICT, online and traditional media

People may communicate face-to-face or through different media channels. Media are formal information channels and may be divided into offline media (without the application of the Internet and its services) and online media (provided by the Internet). Some media exist in both media types, e.g., television. Offline media are mass media (books, newspaper, and magazines) and means for individual information transmission (landline telephone, SMS, letters, and postcards). To use online media presupposes the application of a device of information and communications technology. Such ICT devices are personal computers, notebooks, laptops, tablets, and smartphones.

Three kinds of online media, namely instant messaging services (messengers like WhatsApp and Skype), social media, and other online media are distinguished. Instant messaging services enable "near-synchronous computer-based one-on-one communication" (Nardi et al., 2000, p. 80). In addition to transmissions of text and image messages, there are also offers that provide video calling functions. Social media are "a group of Internet-based applications which are built on the ideological and technological foundations of Web 2.0 and enable the creation and exchange of user-generated content" (Kaplan & Haenlein, 2010, p. 61). Social media can be categorized by functionality (Fischer et al., 2020) into social network services (e.g., Facebook, LinkedIn), self-presentation services (e.g., Instagram, TikTok, Snapchat), micro-blogging services (e.g., Twitter), live streaming services (e.g., YouNow, Twitch), and entertainment services (e.g., Pinterest, YouTube).

Other online services with interest for our study are e-mails, translation services, search engines, encyclopedias, learning services, news websites, digital games, podcasts, dating services, and, finally, streaming services (e.g., Netflix).

6.4 Intuitive research model

Asylum seekers are at the center of this study. The research model integrates the role of asylum seekers and combines it with the present concept of information behavior, the media, and the Uses and Gratifications approach (Figure 6.1). The asylees' information behavior and their media usage for all participants and, additionally, their gender, age (Haji et al., 2020; Zimmer & Scheibe, 2020), language skills, and information literacy skills are described. As learnt from the extended Lasswell formula, the circumstances of media use are important. So, the focus is on the asylees' situation and their information horizons in the former home country and the target country (i.e., Germany). The aspects of the situations during forced migration or in a refugee camp are more on the margins of the investigation and were only considered as part of the literature review.

If there is an information need – and it can be assumed that there are lots of situations in a new country asking for information – it will be satisfied by information production and information reception. For all information needs, analyzed are the motives for producing or receiving information, which are given by the Uses and Gratifications Theory, namely looking for or giving knowledge, seeking entertainment, finding social interaction and socializing with other people, and, finally, trying to present oneself or to establish one's personal identity.

Satisfying an information need means to use media or to talk face-to-face to other people. Concerning the media, it is distinguished between offline media, social media (Scheibe et al., 2019), messengers, and other online media. As the ICT use is a presupposition of the application of digital media, described are the ICT devices applied by asylum seekers. Additionally, they were asked about the concrete applied services and the specific content. For every applied service (e.g., Facebook, search engines, or Twitter), the intensity of usage (e.g., daily, weekly) for the satisfaction of the motives defined by the U> are analyzed. In the same way, the content (e.g., news on the home country, news on Germany, jobs, law, health, education) for every motive and every information service was asked about. Concerning the information horizons, this was asked about twice, for the asylees' information horizon in the former home country and their horizon in the target country.

Given the intuitive research model, formulated are concrete research questions (RQs), which are answered in the next chapters:

Chapter 8
RQ1: How do asylum seekers use ICT, online and traditional media as well as social media for integration according to the literature?

Chapter 9
RQ2a: What ICT, online and traditional media as well as social media do asylum seekers use? Can age- and gender-dependent results be observed?
RQ2b: How do asylum seekers perceive their skills to accurately use ICT?
RQ2c: What ICT, online and traditional media as well as social media do asylum seekers apply to satisfy their needs for information, entertainment, social interaction, and self-presentation? Can age- and gender-dependent results be observed?

Chapter 10
RQ3a: What ICT, online and traditional media as well as social media did asylum seekers use in their former home country and in Germany?
RQ3b: How frequently do or did asylum seekers use ICT, online and traditional media as well as social media in their former home country and in Germany?
RQ3c: What were the asylum seekers' motives in their former home country and in Germany to apply ICT, online and traditional media as well as social media?

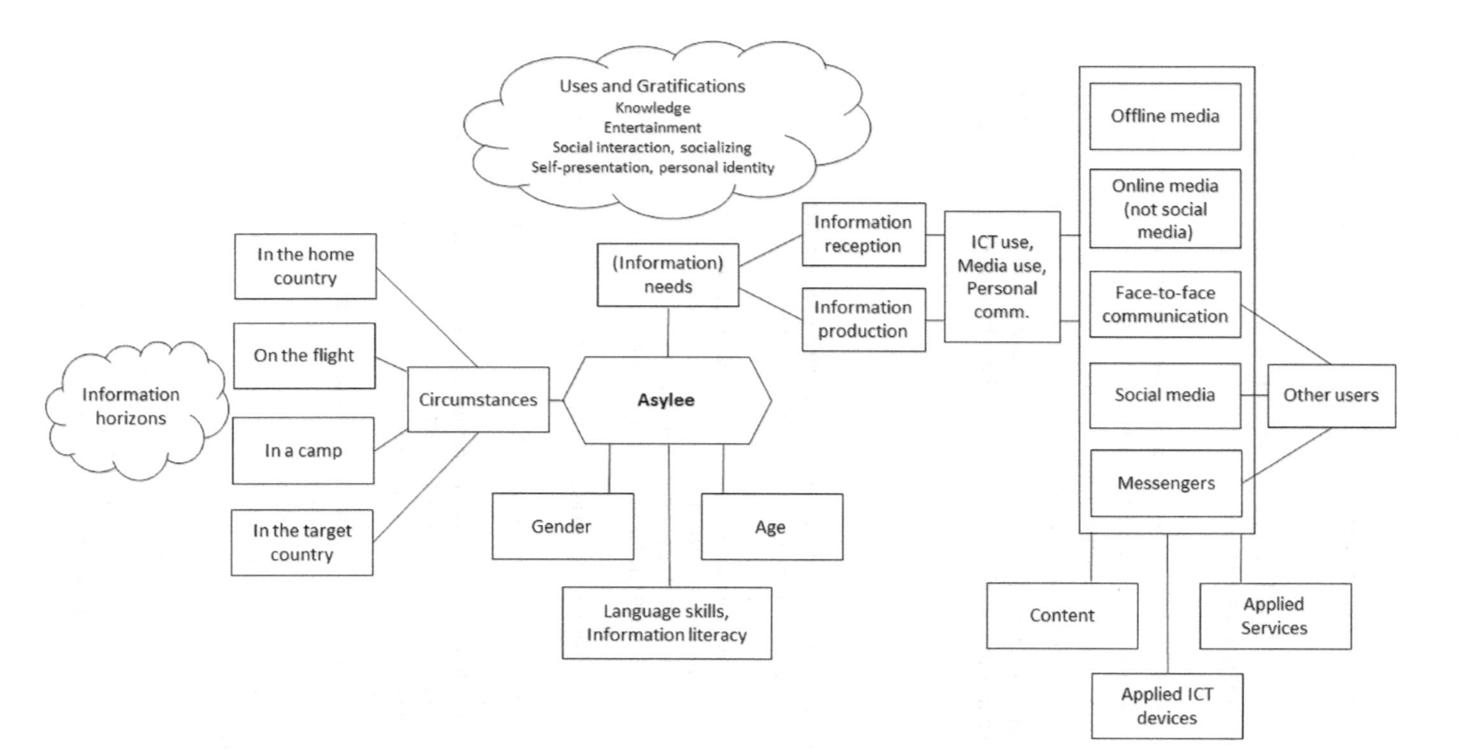

Figure 6.1: Information behavior and media use of asylum seekers.

Chapter 11
RQ4: What kind of information do asylum seekers search for in online forums?

Chapter 12
RQ5: What are experts' insights on the information behavior as well as ICT and media usage of asylum seekers?

The individual research questions contribute to answering what information behavior asylum seekers express and how their information behavior and media usage differs in the target country compared to their former home country. In order to collect data and information for answering the research questions, a multiple methods approach was applied. Different qualitative and quantitative methods were used, which are described in more detail in the following chapter (chapter 7).

References

ACRL. (2000). *Information literacy competency standards for higher education*. American Library Association / Association for College & Research Libraries.

Anfara Jr., V. A., & Mertz, N. T. (2015). Setting the stage. In Anfara Jr, V. A., & Mertz, N. T. (Eds.), *Theoretical frameworks in qualitative research* (pp. 1–20). Sage.

Bates, M. J. (2002). Toward an integrated model of information seeking and searching. *New Review of Information Behaviour Research*, *3*, 1–16.

Blumler, J. G., & Katz, E., Eds. (1974). *The uses of mass communications: Current perspectives on gratifications research*. Sage.

Braddock, R. (1958). An extension of the 'Lasswell Formula'. *Journal of Communication*, *8*(2), 88–93.

Case, D. O. (2007). *Looking for information. A survey of research on information seeking, needs, and behavior* (2nd ed.). Elsevier.

Case, D. O., & Given, L. M. (2018). *Looking for information: A survey of research on information seeking, needs, and behavior* (4th ed.). Emerald.

Cole, C. (2012). *Information need. A theory connecting information search to knowledge formation*. Information Today.

Fischer, J., Knapp, D., Nguyen, B. C., Richter, D., Shutsko, A., Stoppe, M., Williams, K., Ilhan, A., & Stock, W. G. (2020). Clustering social media and messengers by functionality. *Journal of Information Science Theory and Practice*, *8*(4), 6–19.

Fisher, K. E., & Julien, H. (2009). Information behavior. *Annual Review of Information Science and Technology*, *43*(1), 1–73.

Greenberg, B. S. (1974). Gratifications of television viewing and their correlates for British children. In J. G. Blumler & E. Katz (Eds.), *The uses of mass communications: Current perspectives on gratifications research* (pp. 71–92). Sage.

Haji, R., Scheibe, K., & Zimmer, F. (2020). Das Informationsverhalten von jugendlichen Asylbewerbern in Deutschland. *Information – Wissenschaft & Praxis*, *71*(4), 216–226.

Kaplan, A. M., & Haenlein, M. (2010). Users of the world, unite! The challenges and opportunities of social media. *Business Horizons*, *51*(1), 59–68.
Katz, E., Blumler, J. G., & Gurevitch, M. (1973). Uses and gratifications research. *Public Opinion Quarterly*, *37*(4), 509–523.
Katz, E., & Foulkes, D. (1962). On the use of the mass media as “escape”: Clarification of a concept. *Public Opinion Quarterly*, *26*(3), 377–388.
Katz, E., Gurevitch, M., & Haas, H. (1973). On the use of the mass media for important things. *American Sociological Review*, *38*, 164–181.
Lasswell, H. D. (1948). The structure and function of communication in society. In L. Bryson (Ed.), *The communication of ideas* (pp. 37–51). Harper & Brothers.
Maslow, A. H. (1954). The instinctoid nature of basic needs. *Journal of Personality*, *22*, 326–347.
McQuail, D. (1983). *Mass communication theory*. Sage.
McQuail, D., Blumler, J. G., & Brown, J. R. (1972). The television audience: A revised perspective. In D. McQuail (Ed.), *Sociology of mass communications* (pp. 135–165). Penguin.
Nardi, B. A., Whittaker, S., & Bradner, E. (2000). Interaction and outeraction: Instant messaging in action. In W. Kellogg & S. Whittaker (Eds.), *CSCW’00: Proceedings of the 2000 ACM Conference on Computer Supported Cooperative Work* (pp. 79–88). Association for Computing Machinery.
Paisley, W. J. (1968). Information needs and uses. *Annual Review of Information Science and Technology*, *3*, 1–30.
Palmgreen, P., Wenner, L. A., & Rayburn, J. D. (1980). Relations between gratifications sought and obtained: A study of television news. *Communication Research*, *7*(2), 161–192.
Pettigrew, K. E., Fidel, R., & Bruce, H. (2001). Conceptual frameworks in information behavior. *Annual Review of Information Science and Technology*, *35*, 43–78.
Savolainen, R. (2007). Information behavior and information practice: Reviewing the “umbrella concepts” of information-seeking studies. *The Library Quarterly: Information, Community, Policy*, *77*(2), 109–132.
Scheibe, K., Zimmer, F., & Stock, W. G. (2019). Social media usage of asylum seekers in Germany. In W. Popma & S. Francis (Eds.), *Proceedings of the 6th European Conference on Social Media* (pp. 263–272). Academic Conferences and Publishing International.
Shao, G. (2009). Understanding the appeal of user-generated media: a uses and gratification perspective. *Internet Research*, *19*(1), 7–25.
Sonnenwald, D. H. (1999). Evolving perspectives of human behavior: contexts, situations, social networks and information horizons. In T. Wilson & D. Allen (Eds.), *Exploring the contexts of information behaviour* (pp. 176–190). Taylor Graham.
Sonnenwald, D. H., & Iivonen, M. (1999). An integrated human information behavior research framework for information studies. *Library and Information Science Research*, *21*(4), 429–457.
Spink, A., & Cole, C. (2004). A human information behavior approach to the philosophy of information. *Library Trends*, *52*(3), 373–380.
Stock, W. G., & Stock, M. (2013). *Handbook of information science*. DeGruyter.
Sundar, S. S., & Limperos, A. M. (2013). Uses and grats 2.0: New gratifications for new media. *Journal of Broadcasting & Electronic Media*, *57*(4), 504–525.
UNESCO & IFLA (2005). *Beacons of the information society. The Alexandria Proclamation on information literacy and lifelong learning*. Bibliotheca Alexandrina.

Wilson, T. D. (1981). On user studies and information needs. *Journal of Documentation*, *37*(1), 3–15.
Wilson, T. D. (1997). Information behaviour: An interdisciplinary perspective. *Information Processing & Management*, *33*(4), 551–572.
Wilson, T. D. (2000). Human information behavior. *Informing Science*, *3*(2), 49–56.
Zimmer, F., & Scheibe, K. (2020). Age- and gender-dependent differences of asylum seekers' information behavior and online media usage. In *Proceedings of the 53rd Hawaii International Conference on System Sciences* (pp. 2398–2407). ScholarSpace.
Zimmer, F., Scheibe, K., & Stock, W. G. (2018). A model for information behavior research on social live streaming services (SLSSs). In G. Meiselwitz (Ed.), *Lecture notes in computer science: Vol. 10914. Social Computing and Social Media. Technologies and Analytics* (pp. 429–448). Springer.

7 Methods

This study is based on an interdisciplinary approach and uses methods that are established in information science and social sciences (Connaway & Radford, 2016). The multiple methods approach (Morse, 2010) was applied to analyze the information behavior of asylum seekers in Germany. The information needs of asylum seekers may shift due to the novel situation they are facing in a new country. This work also examines which information sources are preferred by asylum seekers while keeping in mind contextual difficulties, e.g., cultural differences and language problems. Some media and ICT that were used in their home country may no longer be accessible and may become unattractive, too time consuming, or simply will no longer be needed. Different and previously not used devices and services could potentially be needed to cope with life and various situations in a new country. To be able to study the information behavior of asylum seekers in their former home and in the new country, they have to be asked about what social media, ICTs, and devices they used to be informed and what information channels were used to exchange information with other people. Additional data was collected through content produced by asylum seekers on information channels. The results for this book are not chronologically ordered since the flow of information seemed more fitting as displayed.

Based on publications by colleagues, a few key themes were adapted for this book. Research on the application of (social) media and ICT prior and during migration served as a first starting point. A study by Dekker et al. (2018) found how media are used as decision tools during forced migration. The findings of the study indicate that refugees mainly trust information from well-known online communities as well as existing social ties. Furthermore, the Internet is mostly accessed by smartphones. Emmer, Richter, and Kunst (2016) investigated how asylum seekers inform themselves in their home country prior to escape, during the flight, and when arriving in Germany. Furthermore, they concentrated on which media they use and which sources asylum seekers trust. Similar to the findings of Dekker et al. (2018), they found that for asylum seekers, mobile phones serve as a multifunctional tool and the Internet as well as the television are very important ICTs in their life. Social media is mainly used for communication. This research mainly concentrates on information behavior and media as well as ICT usage for integration. The study investigates the purpose of social media channels during the integration process. Results indicate that various services are used by asylum seekers (e.g., WhatsApp, Facebook, and YouTube) several times a day and that social media is a very essential and relevant aspect in the integration process as it is used to connect with other people, to learn about the new culture, and to improve

https://doi.org/10.1515/9783110672022-007

language competences. Furthermore, asylum seekers were able to build bonding and bridging social capital. Alencar and Tsagkroni (2019) explained how refugees have to have an active role during the two-way integration process. Here, it becomes obvious how digital media and ICTs act as a mediator in this process. Through media, networks can be built, traditional as well as online, to access information for self-support. In this scope, it is argued that online content should be made more inclusive to target the needs of various asylum seeker populations.

Concentrating on the different aspects of information behavior, it is defined as a concept that includes information production, information seeking, information use, and furthermore information needs (Wilson, 2000). With information seeking, one means not only these activities which contain the active search for information, but also situations in which information is consumed passively (Bates, 2002; Wilson, 1997; Stock & Stock, 2013). For example, the passive consumption of information happens during human interaction, resulting in information exchange (Wilson, 1981). Information behavior depends on context and situation (Dervin, 1997) and on one's personal information horizon (Sonnenwald, 1999).

For the theoretical foundation in regards to media research, the interview as well as the survey questionnaires are based on the Uses and Gratifications Theory (U>) by Katz et al. (1973). It is a popular approach in mass communication studies to explain motives for media use and identify what kind of needs are satisfied. Later, McQuail (1983) summarizes four central motives for media use of different needs proposed by Katz et al. (1973), which are: information, entertainment, social interaction, and self-identity. Adapted from Shao (2009) who differentiates between three different user roles: consumers, participants, and producers, the fourth motivation, self-identity, transformed to self-presentation as a motivation for using media. Here, the users who actively produce and share content online do so as a form of self-expression and self-actualization.

7.1 Multiple methods approach: Gathering qualitative and quantitative data

To gather the necessary data and information, a multiple method investigation was conducted by combining qualitative and quantitative approaches. For qualitative data, semi-structured face-to-face interviews were performed and quantitative data were collected with questionnaires. Furthermore, a content analysis served to gather further qualitative data.

The goal of this study is the analysis of the interaction between individuals and information, as the field of information science proposes. As this is a complex concept, the application of different methods enables thorough insights combining quantitative and qualitative information. Yet, according to Togia and Malliari (2017) as well as Fidel (2008), a mixing of methods is a rare approach in information science research. In contrast, Hildreth and Aytac (2007) see a rising number of applications of this procedure. Nonetheless, the combination of methods fosters a deeper understanding of a topic and minimizes the weaknesses each of the quantitative and qualitative approaches possess (Togia & Malliari, 2017), making this approach more popular in social sciences (Poteete et al., 2010). As with social sciences, case studies and small-N research designs are helpful in theory development and the analysis of hypothesized causal sequences. But, of course, small N-studies cannot offer an evaluation of general relationships (Poteete et al., 2010).

The multiple method research as well as mixed methods approach are defined through the application of both qualitative and quantitative methods or a mixture of both types of methods. Therefore, any research with different types of analysis, approaches or data, and also two different study populations can be considered (Morse, 2010). The research questions can be addressed by employing various combinations of field, survey, experimental, or other methods (Brewer & Hunter, 2005).

Differentiating the multiple methods approach from the mixed methods approach, the multi method approach consists of multiple applied methods, whereby each applied research method may be published as a standalone research study and does not necessarily need the other methods (Morse, 2010). In contrast, the mixed method design focuses on one core project applying one complete method, as well as a second project employing a different type of method or analysis (Cresswell et al., 2003). Here, one method alone would be incomplete, as a publication of the second project and data would not be possible without the core project. The advantage of this approach is that another area which is pertinent to the research question of the core project cannot be included in the main research project but through separating it as an extension of the study, multiple data points can be analyzed. The data is collected until the researchers are certain they can answer this special part of the research question (Morse, 2010).

The multiple methods design can include two or more studies concerned with the same research question(s) or different parts of the same project or question. The advantage of this method is that there are no methodological issues as each project is publishable on its own. It is not uncommon to conduct projects and publish them separately for the synthesization of the findings to illustrate the relationships in regards to the data for answering the overall research question (Morse, 2010).

To sum up, with the help of the multiple methods approach, it is possible to collect a larger amount and variety of data that can later be combined to answer the research questions, as "no single method overcomes all challenges" or "fully addresses all standards" (Poteete et al., 2010, pp. 5, 11). As a starting point we conducted desk research by reading on various literature about asylum seekers' media practices to gain first insight into the topic. For our approach, qualitative and quantitative data about the information behavior and online, traditional, and social media usage of asylum seekers related to the Uses and Gratifications Theory were collected with a survey (survey 1) and furthermore by interviewing asylum seekers, adults and children (paragraph 7.3). The data of another survey (survey 2) represent the differences of asylum seekers (social) media usage while living in their home country and after arriving in Germany (paragraph 7.3). A content analysis (paragraph 7.4) of a forum was applied to gather qualitative data about asylum seekers' information needs. By interviewing experts (paragraph 7.5), further qualitative data about topics related to the information needs, information behavior, and information literacy as well as to the ICT and social media, and traditional and online media usage by asylum seekers was collected. By conducting a literature review (paragraph 7.2), the researchers get insights about existing literature and results from other research studies (Figure 7.1).

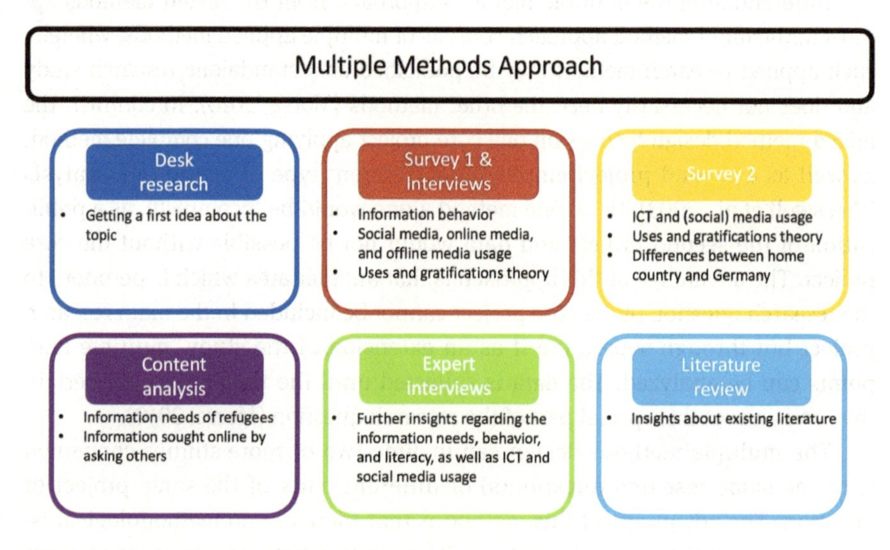

Figure 7.1: Overview of applied multiple methods.

7.2 Literature review: Identifying literature on ICT and media practices for integration

In order to identify the information behavior and needs of newly arrived and, additionally, asylum seekers living for a longer amount of time in a new country, a literature review was conducted. The approach for the literature review was based on Fink (2019) and Moher et al. (2009). The databases used to retrieve the documents for this study were Web of Science and Scopus, as they are commonly applied for literature review and indicate a certain qualitative standard. These retrieval systems are multidisciplinary databases, meaning they cover wide areas of research. Both can be used to search for, e.g., topics, titles, abstracts, keywords, and authors. As the databases differ in the amount of journals and proceedings, it is recommended to use more than one information service to enhance the overall recall (Fink, 2019).

To add the literature to the potential investigation corpora, a few requirements needed to be met. The search, conducted on June 29, 2020, consisted of a search query dealing with the keywords "refugee," "asylum seeker," "asylee," "forced migration," "media," "internet use," "ict," "digital," "communication," "information behavior," "social media," "uses and gratifications." The corresponding search queries are:

Web of Science
TS = (("refugee*" OR "asylum seeker*" OR "asylee*" OR "forced migration") AND ("uses and gratifications" OR "media" OR "Internet Use" OR "ict" OR "digital" OR "communication*" OR "information behavior" OR "social media"))

Scopus
TITLE-ABS-KEY (("refugee*" OR "asylum seeker*" OR "asylee*" OR "forced migration") AND ("uses and gratifications" OR "media" OR "Internet Use" OR "ict" OR "digital" OR "communication*" OR "information behavior" OR "social media"))

Furthermore, the timeframe from 2015 to June 2020 was chosen in order to gain insight into the information behavior of asylum seekers corresponding to our research focus to establish a more current and comparable approach with our data. In addition, papers published in journals or conference proceedings as well as book chapters were included, excluding workshops and posters. This resulted in a corpus consisting of 2,575 papers. Further, potential literature seen as relevant was extracted from the references of the chosen literature and added to the potential investigation corpus for further examination, adding 156 articles. All papers considered in this review were written in German or English. Several rounds to review the literature were conducted, narrowing down on the scope of the relevant literature, as can be seen in the PRISMA flow diagram according to Moher et al. (2009) (Figure 7.2).

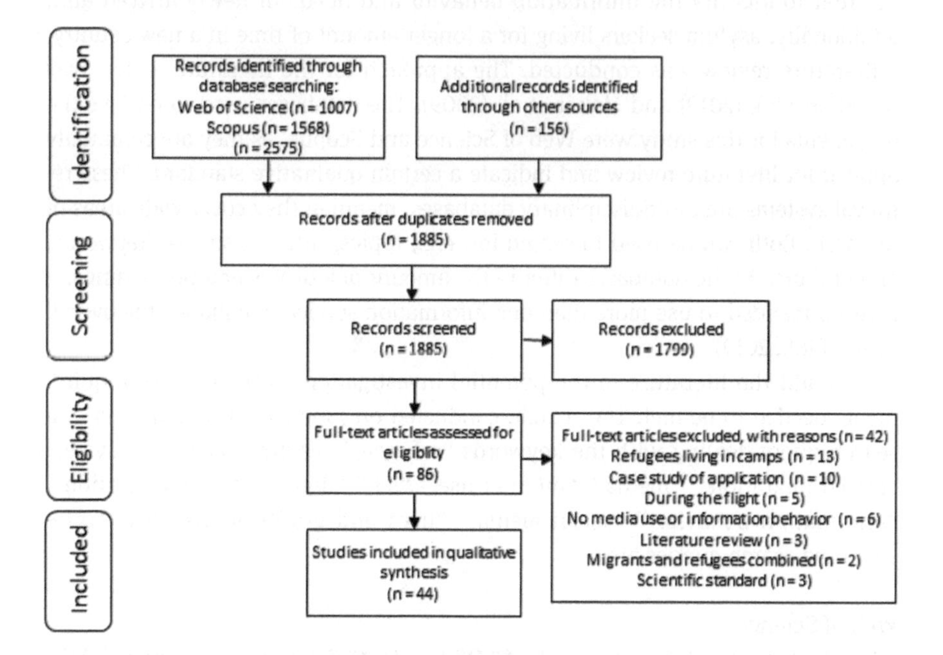

Figure 7.2: PRISMA flow diagram. Adapted from Moher et al. (2009).

To be included for the subsequent content analysis, the studies needed to investigate asylum seekers who live in a new country and therefore excluded studies examining asylum seekers during forced migration or while living in refugee camps. As meta reviews and literature reviews are considered secondary data, these articles were not included in our literature review. A large fraction of studies also investigates media coverage and related topics of asylum seekers which were also excluded, as this study is solely interested in the usage of media by asylum seekers. Furthermore, the literature was chosen if information behavior according to Wilson (1997), Pettigrew et al. (2001) as well as Stock and Stock (2013) could be observed. This included mentioning of specific information needs or the distribution of information, often dealing with the application of online, traditional or social media, as well as several ICT. Similar to the information dimension, the other three motivations of the U> were applicable as follows. The socialization dimension was seen as suitable every time the media was used to get in contact with other people (e.g., family, friends, and new contacts). This could be, for example, if a study explained how the participants mentioned that they use smartphones to contact relatives. Concerning the entertainment motive, it was applicable if the media was used for relaxation, fun, or similar purposes. Self-presentation is expressed through posting pictures, videos, status (via e.g., WhatsApp), and similar content on social media. This also includes postings of things related to oneself (e.g., selfies, pictures of the family, or poems).

Consequently, the literature was analyzed and concepts identified to gather more expansive knowledge on the discussed themes. The information was intellectually extracted from the relevant studies (N=44). The codes used for the analysis included: information (news, health, education, employment, etc.), entertainment, socialization, self-presentation, social media (Facebook, WhatsApp, Snapchat, etc.), ICT (smartphone, computer, TV, etc.), media (newspaper, online news websites, etc.), nationality (Syrian, Afghan, Eritrean, etc.), country of study (Germany, Sweden, etc.). Each paper was categorized according to the following topics and sub-themes:

- ICT and media usage in a new home country
- problems related to ICT and media usage
- information: language learning, education and employment, health, law, news, everyday information and tasks
- socialization: keeping old contacts, new contacts, problems related to ICT and social media usage with socialization
- entertainment
- self-presentation
- information exchange

Furthermore, the review was partially summarized in tabular form and described in a short summary. The identified and synthesized literature and their characteristics will be described in chapter 8 identifying the information needs and information behavior of asylum seekers integrating into a new society. Only prominent papers have been chosen by the research group to be investigated deeply.

7.3 Interviews and surveys: ICT and media practices in a new country

To gather data, a combined quantitative and qualitative approach was employed (Sonnenwald & Iivonen, 1999), namely the interview methodology and the survey methodology, i.e. “a sample of individuals is asked to respond to questions” (Case & Given, 2018, p. 236). Due to experienced difficulties to reach asylum seekers through information channels such as social media and e-mail, various institutions and organizations were contacted. The contact addresses were received from the Federal Office for Migration and Refugees (*Bundesamt für Migration und Flüchtlinge*) or by researching further administrative addresses involved in integration processes.

Concentrating on different key topics, two phases of interviews and surveys were conducted. The first phase (phase 1) concentrated on a combined approach of interviews and a corresponding survey, the second phase (phase 2) employed the distribution of a survey.

For the combination of the survey and the interviews (phase 1), different associations, institutions, and organizations as well as city governments in North-Rhine Westphalia were approached as early contact points. It was recommended to interview asylum seekers attending German language classes in order to properly communicate. The interviews were conducted in Dorsten, Düsseldorf, Wesel, and Moers. Overall, two rounds of interviews for phase 1, all held in either German or English, were performed. The authors spoke to the interview partners in person. While one researcher interviewed the participant, the other filled out a corresponding survey. As research with refugees and asylum seekers naturally presents ethical issues due to the vulnerability of those being interviewed, we refrained from speculating about personal stories or histories of our interview partners. Further, we did not include questions about legal aspects of the person's asylum procedures or political issues to not jeopardize the interview partner's feelings (Borkert et al., 2018). All interview partners were informed about the nature of this study – the data will only be used for scientific publications. All interviews were transcribed by the two researchers. For this book, the texts had to be translated from German to English, whereby most sentences are reproduced analogously to their meaning because of the language difficulties.

The semi-structured interview-guide and corresponding survey were designed to analyze the information and communication behavior of asylum seekers. Therefore, the gratifications sought and obtained have to be considered as embedded in a special context (being a migrant in a foreign country) and a very special situation (arriving in the country, being not allowed to work in Germany for the first time after arrival, and not always speaking the German language fluently). Indeed, as Case and Given (2018) state, "context and situation are important concepts for information behavior research" (p. 48). We can assume that the migrants' information horizons are very special and are influenced by the situation in the home country, the migration from their home country to Germany, and the unfamiliar circumstances along with new social contacts in Germany.

The second phase (phase 2) encompassed an additional survey which is based on the results from phase 1. Its main purpose was to detect behavioral changes in the media usage and the information behavior differentiating between the asylum seeker's home country and Germany. The survey was distributed in German language courses in Düsseldorf and was available in German, whereby the results had to be translated to English for this research. The researchers visited the language classes and were available for questions. The participants could fill out the questionnaire online, and additionally, the survey was printed out and handed to participants who did not want to use the online version. After completing the survey, the sheets were collected by the researchers. Participants had to select the social media services, ICTs, online media, and traditional media they were using in their home country or are using now in Germany. The duration of the media usage for the home country and for Germany were stated on a Likert-scale. Furthermore, the attendees were asked for their motives to use the media.

Both questionnaires as well as the interview guides from phase 1 and phase 2 focused on the asylum seekers' ICT and (social) media service usage in relation to the Uses and Gratifications Theory (Katz et al., 1973; Blumler & Katz, 1974) and the summarized four central motives to use media (McQuail, 1983), which are information, entertainment, social interaction, and self-actualization. Self-actualization was transformed to self-presentation for producing content (Shao, 2009).

Interviews: Adults

The first round of interviews took place in two separate education centers (*Volkshochschulen*). The researchers visited the first education center on November 28 and 30 as well as December 3, 2018, where 19 participants were interviewed. The second interview location was visited in February 2019 where teachers selected students for the interviews, resulting in six additional interview partners. All interview participants were attendees of German language courses of different levels (A1: beginner to C1: proficiency), who voluntarily took part in the interviews. For asylum seekers who were not fluent in German yet, an Arabic native speaker was present for the interviews in the first interview location. For the ones conducted in the second education center, the language skills of the interview partners were sufficient. The adults were aged between 21 and 55. Overall, the interviews took around 15 to 30 minutes each and were recorded with a recording device.

First, the interviewees were asked about demographics, country of origin, their age, gender, educational background, and German language level. Then, the participants were asked to state what kind of ICT (smartphone, Internet, TV, laptop, tablet, PC, radio, landline) they use and how they would rate their ICT and (online) media skills on a five-point Likert-scale (Likert, 1932), 1) meaning "very bad," 3) "neither bad nor good," or 5) "very good" skills.

The interviewers asked about specific online or traditional media, as well as social media and messaging service usage for each of the four main gratifications (information, entertainment, social interaction, self-presentation). Social media services and messaging services included: Facebook, Instagram, live streaming services (e.g., Periscope, Twitch), Twitter, TikTok, Snapchat, WhatsApp, YouTube. Interviewees always had the option to name additional services. Concerning the different motivations and gratifications, an overview of the asked media and ICT are as follows.

- Information: social media and messaging services (see above), news websites, online encyclopedias (e.g., Wikipedia), newspapers, magazines, books, radio, and TV. Concerning the information content, we asked for news, employment, education, law, health, and religion. In each case we provided a category "other;"
- Entertainment: social media and messaging services (see above), gaming apps, streaming services (e.g., Netflix), newspapers, magazines, books, radio, TV;
- Social interaction: social media or messaging services (see above), SMS, landline telephone, Skype;
- Self-presentation: social media or messaging services (see above).

Interviews: Children

After a first analysis of the collected data, a second round of semi-structured interviews and filling in questionnaires was performed with 21 children and adolescents aged between eight to 18 years in April 2019. The procedure took place during a holiday care program for children of asylum seekers. Because of the age and attention span of the children, the interviews took around 15 minutes each in which they could answer questions during a personal conversation and talk about their individual experiences. If communication difficulties occurred, the children helped each other or volunteers working for the program were able to translate.

As the children have potentially different needs than the adult interview partners, the semi-structured interview guide was adapted: The children were asked about demographic data and their dream job to start with something simple and relatable. Afterwards, questions related to the children's ICT and traditional and online media usage were asked. ICT and further online and traditional media they use since living in Germany was asked about as well. All media and ICT were adapted from the first interview round. Afterwards, the children could rate their ability to handle the media and digital devices on a five-point Likert-scale (similar to the adults). The children were asked about their motives for producing and consuming content, i.e., about information, entertainment, social interaction, and self-presentation. Here, both topics (e.g., homework, learning) and information services (such as Twitter, Facebook, WhatsApp, YouTube, Instagram, or TikTok) were given, whereby it was also possible to state additional services. Displayed in Tables 7.1 and 7.2 are the demographic data of the interview and survey participants. In all cases, names have been removed for privacy reasons.

Table 7.1: Demographics of interview partners (adults).

#	Age	Gender	Country	Native language	Educational level	Months in Germany
IP1	34	male	Afghanistan	Farsi	abitur (A level; university entrance qualification)	24
IP2	35	female	Syria	Kurdish	school until age 6	36
IP3	36	female	Syria	Kurdish, Arabic	university (fashion design)	36
IP4	37	male	Iran	Kurdish, Farsi	abitur	24
IP5	34	female	Syria	Arabic	school until 9th grade	36
IP6	24	male	Syria	Arabic	school until 10th grade	36
IP7	38	male	Syria	Arabic	9 years in school	42
IP8	37	male	Syria	Arabic	until intermediate level	38
IP9	30	male	Iraq	Arabic	abitur	36
IP10	27	female	Syria	Arabic, Kurdish	university	36
IP11	34	female	Syria	Arabic	university (educational studies)	48
IP12	30	female	Syria	Arabic	university (french studies)	12
IP13	28	male	Syria	Arabic	abitur	30
IP14	40	female	Syria	Arabic	university (agricultural engineer)	24
IP16	27	male	Syria	Arabic	university (legal studies)	36
IP17	21	male	Afghanistan	Dari	school until 8th grade	36
IP18	28	male	Afghanistan	Dari	school until 4th grade	36
IP19	55	female	Iran	Farsi	abitur	48
IP20	29	male	Syria	Arabic, Kurdish	university (economic studies)	36
IP21	26	male	Syria	Arabic	university (media studies)	36

Table 7.1 (continued)

#	Age	Gender	Country	Native language	Educational level	Months in Germany
IP22	22	female	Turkey	Turkish	university (legal studies)	15
IP23	26	male	Turkey	Turkish	university (business administration)	12
IP24	25	female	Syria	Arabic	abitur	42
IP25	40	male	Syria	Arabic	university (legal studies)	41

Table 7.2: Demographics of interview partners (children).

#	Age	Gender	Country	Native language	Educational level	Months in Germany
IP26	14	male	Syria	Arabic	in 8th grade	24
IP28	16	female	Syria	Arabic	in 8th grade	24
IP29	12	female	Syria	Arabic	in 5th grade	24
IP30	12	male	Azerbaijan	Azerbaijani	in 6th grade	48
IP31	14	male	Turkey	Turkish	in 9th grade	4
IP32	12	female	Turkey	Turkish	in 5th grade	4
IP33	13	female	Afghanistan	Farsi, Dari	in 7th grade	36
IP34	16	male	Afghanistan	Dari	in 8th grade	36
IP35	16	female	Afghanistan	Dari	vocational college	24
IP36	13	male	Afghanistan	Dari	in 6th grade	24
IP37	18	male	Afghanistan	Dari	in 9th grade	16
IP38	12	male	Afghanistan	Dari	in 4th grade	24
IP39	11	male	Lebanon	Arabic	in 4th grade	37
IP40	9	male	Iraq	Arabic	in 5th grade	72
IP41	10	female	Syria	Kurdish	in 4th grade	24
IP42	9	female	Syria	Kurdish	in 3rd grade	48
IP43	10	male	Syria	Arabic	in 4th grade	5
IP44	11	male	Iraq	Kurdish	in 3rd grade	48

Table 7.2 (continued)

#	Age	Gender	Country	Native language	Educational level	Months in Germany
IP45	9	male	Azerbaijan	Azerbaijani	in 2nd grade	48
IP46	8	male	Turkey	Turkish	in 2nd grade	72
IP47	9	male	Iraq	Kurdish	in 2nd grade	48

(Online) survey

Since information behavior depends on the given circumstances and conditions (Case & Given, 2018), it is important to study whether there are behavioral changes in the behavior of asylum seekers before they leave their home country and after arriving at their final destination, concerning our participants, Germany. A sample of the target group was asked with the help of a questionnaire to gather quantitative data (Fowler, 2013).

Based on open questions and results from the interview and survey round, the second survey was prepared (phase 2). First, the online survey was available as a pre-test from June to September 2019 and a total of 17 asylum applicants participated. It was distributed through SurveyCircle, Facebook, and contacts with German language school institutions. After the pre-testing phase, the survey was reconstructed and adapted. Therefore, the results from the 17 participants are not represented in this study. To distribute the newly prepared survey, the researchers visited an integration language school in Germany on February 12, 2020 and February 14, 2020 and visited language classes with a B2 language level or higher. Furthermore, the researchers and language teachers had the opportunity to support the participants while answering the survey questions.

The questionnaire was available as an online survey on umfrageonline.com. Alternatively, the survey was handed out as a print version for participants who did not want to use the online version. Concerning the structure of the survey, first, the participants were asked if they were using a certain kind of social media platform, ICT, traditional media, or online media in their home country and, furthermore, which kind they use in Germany. Therefore, it is possible to identify the changes of the behavior prior to escaping the home country and after arriving in Germany.

To investigate the used ICTs and social media as well as online or traditional/offline media, the following options were given:

- Social media: Facebook, Instagram, LinkedIn, live streaming services (e.g., Periscope, Twitch), messaging services (e.g., WhatsApp, Facebook Messenger, Telegram, Viber), Pinterest, Skype, Snapchat, TikTok, Twitter, YouTube, dating apps (e.g., Tinder, Badoo, Lovoo);
- ICT: Computer/notebook/laptop, television, landline telephone, smartphone, tablet;
- Traditional/offline media: Books, letters/postcards, newspapers/magazines, SMS;
- Online media: Digital games (e.g., PUBG, Candy Crush), e-mail, encyclopedia (e.g., Wikipedia, Encyclopedia Britannica), learning apps (e.g., Duolingo, Babbel), podcasts, search engines (e.g., Google, Bing), streaming services (e.g., Netflix, Amazon Prime, Sky Go), translation services (e.g., Google Translate, DeepL), websites of news services.

After choosing the appropriate ICTs, social media, and online or traditional/offline media used in the home country and in Germany, the participants were asked how often they used the services in the home country and now in Germany. This could be answered with the help of a seven-point Likert-scale. The options to answer were: 1) "never," 2) "a few times a year," 3) "about once a month," 4) "a few times a month," 5) "about once a week," 6) "a few times a week" as well as 7) "daily." Furthermore, the survey attendees were asked if they used the respective ICT or media for (1) information (knowledge), (2) entertainment, (3) socializing (general) as well as socializing in order to be able to communicate with others, or (4) personal identity and self-presentation from the U>. Again, a seven-point Likert-scale was used with the possibilities to answer from (1) "totally disagree" about (4) "neutral" up to (7) "totally agree." The participants always had the option to answer "I don't know." Lastly, the participants were asked about their demographic data. This included their nationality, their native language, age, gender, level of education, in which year they arrived in Germany, and their German language level. In addition, they were asked whether private contact with Germans is important to them and how often they have private contact with Germans. The questions were asked in German and were translated to English for the manuscripts and the representation of the results.

7.4 Information needs: Content analysis of a forum (Wefugees)

To expand the findings about information needs of asylum applicants during their orientation and integration in another country, we applied a content analysis (Krippendorff, 2004) on questions from refugees and asylum applicants. The

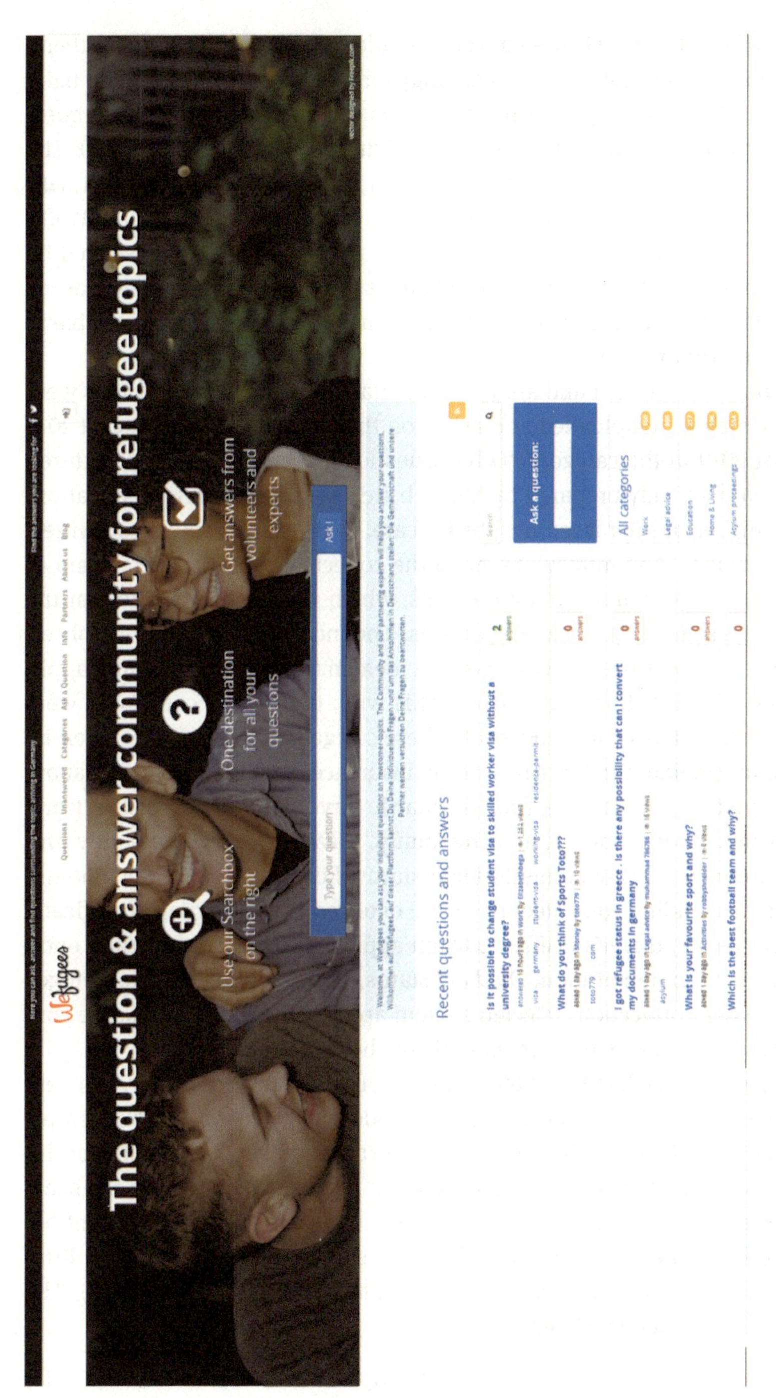

Figure 7.3: Screenshot of Wefugees homepage (September 26, 2020).

questions collected were asked on a website called *Wefugees*, who present themselves as "the question and answer community for refugee topics" (Wefugees, n.d.). Wefugees is an interactive platform where users can ask, find, and answer questions about topics dealing with arriving and living in Germany (Figure 7.3). The goal is to offer asylum seekers the opportunity to be independent and self-sufficient. The platform is supported by aid agencies being selected by the Wefugees team. The questions asked by refugees or asylum seekers are answered by German volunteers who are experienced with asylum seekers information needs and other related concerns. This way, the information can be made accessible to others in a sustainable way.

The data was collected and analyzed manually. In total, 244 questions were collected, which were uploaded and asked on the website between August 2015 and August 2019 in the category "Other Questions," since other pre-structured categories were already prepared by the website (work, legal advice, education, home & living, asylum proceedings, healthcare, information & offers, activities, "How can I help?," and money). Some of the collected questions were spam or not asked by refugees or asylum applicants, which were not considered for the content analysis, resulting in 139 questions. The inductive approach via observation of the data was used to get a general idea on the questions asylum applicants asked (Hsieh & Shannon, 2005). This way, the coding categories were defined and characterized in a codebook. These categories are: Marriage, paternity rights, job, integration into German culture, insurance, supplies, accommodation, language, institution, culture (not German), travel, studying, leisure time, dating, social contacts, media, missing family, travel route, rights, family unification, interview, and documents/legal status. For the category documents/ legal status responding sub-categories were detected: Visa, marriage certificate, ID, birth certificate, driver's license, asylum application, residence permit, entry permission, ban on deportation, *Duldung* status, registration certificate, refugee status, subsidiary protection, rejected asylum application, deportation. The content analysis was conducted independently by the two researchers.

To determine the reliability of the content analysis among the coders, Krippendorff's alpha was calculated. Hayes and Krippendorff (2007) describe reliability as: "it amounts to evaluating whether a coding instrument, serving as common instructions to different observers of the same set of phenomena, yields the same data within a tolerable margin of error. The key to reliability is the agreement observed among independent observers" (p. 78). As our data is binary coded, Krippendorff's alpha seems sufficient for the calculation (Krippendorff, 2004). The general form of Krippendorff's alpha is:

$$\alpha = 1 - D_O / D_e \qquad (1)$$

D_O is the observed disagreement between the coders, whereas D_e is the one predicted by chance. If the coders agree completely, the disagreement D_O is 0 and alpha is 1. If alpha is 0 and D_O equals D_e, then there is no agreement between the coders. According to Krippendorff (2004), alpha should be ≥ .800. After the coding process, Krippendorff's alpha was calculated on nominal data to measure the agreement between the two coders and ensure reliability. The calculation of the Krippendorff's alpha matrix resulted at $\alpha = 0.973$, indicating a great inter-coder reliability (Table 7.3).

Table 7.3: Krippendorff's alpha matrix. 0 and 1 describe value pairs, e.g., 0-0 means both coders agree the document is not relevant, for 1-1 both coders agree it is relevant.

	0	1	Total
0	5,982	10	5,992
1	10	392	402
Total	5,992	402	6,394

7.5 Qualitative expert interviews

With the gained knowledge from the author's previous studies (Haji et al., 2020; Scheibe et al., 2019; Zimmer & Scheibe, 2020), semi-structured interviews with experts whose research articles were observed during the review of related literature to gather further qualitative data were conducted. Semi-structured interviews allow a more flexible variation for the order and use of questions. It provides the possibility to ask further in-depth questions and creates room for explanations and experiences of participants. Overall, a mixture of open-ended and theoretical questions was used (Galletta, 2013).

The findings should be consolidated and further knowledge about experiences regarding asylum seekers' information behavior, their information literacy, as well as used information channels and sources can be collected. This approach was chosen to deepen an understanding of the results gathered through interviews, surveys, and content analysis. As there exists not *the* expert interview methodology, a range from quantitative measures as a form of information source to theoretically, qualitative approaches can be used (Bogner et al., 2009). Of course, experts can be considered as an objective source of information, but this notion has been discussed in scientific research (Bogner et al., 2009). Therefore, the expert interviews were conducted after careful examination of

Table 7.4: Information about the experts.

Reference	Expert	Topic	Affiliation
IE 1	Prof. Dr. Carola Richter	Media use of asylum applicants	Freie Universität Berlin, Germany
IE 2/3	Dr. Juliane Stiller	Information literacy	You, We & Digital, Berlin, Germany
	Dr. Violeta Trkulja	Information literacy	You, We & Digital, Berlin, Germany
IE 4	Prof. Dr. Rianne Dekker	Online and social media in migration networks	Utrecht University, The Netherlands

the studies' results to foster discussion among the researchers rather than being considered as an objective information gathering meeting.

In total, four experts (Table 7.4) were interviewed, whereof two have been interviewed together. The interviews took place on June 29, 2020 (Dr. Juliane Stiller and Dr. Violeta Trkulja), July 3, 2020 (Prof. Dr. Carola Richter), and July 9, 2020 (Prof. Dr. Rianne Dekker) and were performed online via Cisco WebEx. The interview with Prof. Dr. Carola Richter and the interview with Dr. Violeta Trkulja and Dr. Juliane Stiller were held in German. Prof. Dr. Rianne Dekker's interview was held in English. Each of the interviews took approximately 60 minutes. During the interviews, the researchers took notes and the interviews were recorded with the consent of the interviewees. Due to technical issues, the interview with Dr. Violeta Trkulja and Dr. Juliane Stiller could not be recorded and statements are based on the notes taken during the interview.

For the interviews, a collection of 12 pre-formulated questions were prepared about the following topics:
- asylum seeker research
- asylum seekers' information behavior, considering information needs, uses, and requests
- information sources and platforms (e.g. social media, forums) used by asylum seekers
- asylum seekers' (social) media use
- asylum seekers' information and digital literacy
- age- and gender-dependent differences

The interviews were transcribed, whereby each German interview was translated into English by word and sense as best as possible. They were then intellectually analyzed according to different topics extracted from the content.

References

Alencar, A. (2018). Refugee integration and social media: a local and experiential perspective. *Information, Communication & Society, 21*(2), 1588–1603.

Alencar, A., & Tsagkroni, V. (2019). Prospects of refugee integration in the Netherlands: Social capital, information practices and digital media. *Media and Communication, 7*(2), 184–194.

Bates, M. J. (2002). Toward an integrated model of information seeking and searching. *New Review of Information Behaviour Research, 3*, 1–16.

Blumler, J. G., & Katz, Eds. (1974). *The uses of mass communications: current perspectives on gratifications research*. Sage.

Bogner, A., Littig, B., & Menz, W. (2009). Introduction: Expert interviews - An introduction to a new methodological debate. In: A. Bogner, B. Littig, & W. Menz (Eds.), *Interviewing experts: Research methods series* (pp. 1–15). Palgrave Macmillan.

Borkert, M., Fisher, K. E., & Yafi, E. (2018). The best, the worst, and the hardest to find: How people, mobiles, and social media connect migrants in(to) Europe. *Social Media + Society, 4*(1), 1–11.

Brewer, J., & Hunter, A. (2005). *Foundations of multimethod research: Synthesizing styles*. Sage Publications.

Case, D. O., & Given, L. M. (2018). *Looking for information: a survey of research on information seeking, needs, and behavior* (4th ed.). Emerald Publishing Limited.

Connaway, L. S., & Radford, M. L. (2016). *Research methods in library and information science* (6th ed.). Libraries Unlimited.

Cresswell, J. W., Plano-Clark, V. L., Gutmann, M. L., & Hanson, W. E. (2003). Advanced mixed methods research designs. In T. Abbas & C. Teddlie (Eds.), *Handbook of mixed methods in social & behavioral research*. Sage Publications.

Dekker, R., Engbersen, G., Klaver, J., & Vonk, H. (2018). Smart refugees: How Syrian asylum migrants use social media information in migration decision-making. *Social Media + Society Society, 4*(1), 1–11.

Dervin, B. (1997). Given a context by any other name: Methodological tools for taming the unruly beast. In P. Vakkari, R. Savolainen, & B. Dervin (Eds.), *Information Seeking in Context* (pp. 13–38). Taylor Graham.

Emmer, M., Richter, C., & Kunst, M. (2016). *Flucht 2.0: Mediennutzung durch Flüchtlinge vor, während und nach der Flucht* [Escape 2.0. Media usage by refugees before, during and after their escape]. https://www.polsoz.fu-berlin.de/kommwiss/arbeitsstellen/internationale_kommunikation/Media/Flucht-2_0.pdf

Fidel, R. (2008). Are we there yet?: Mixed methods research in library and information science. *Library and Information Science Research, 30*(4), 265–272.

Fink, A. (2019). *Conducting research literature reviews* (5th ed.). Sage Publications.

Fowler, F. J. (2013). *Survey research methods* (5th ed.). Sage Publications.

Galletta, A. (2013). *Mastering the semi-structured interview and beyond: From research design to analysis and publication*. New York University Press.

Haji, R., Scheibe, K., & Zimmer, F. (2020). Das Informationsverhalten von jugendlichen Asylbewerbern in Deutschland. *Information-Wissenschaft und Praxis, 71*(4), 216–226.

Hayes, A. F., & Krippendorff, K. (2007). Answering the call for a standard reliability measure for coding data. *Communication Methods and Measures, 1*(1), 77–89.

Hildreth, C., & Aytac, S. (2007). Recent library practitioner research: A methodological analysis and critique. *Journal of Education for Library and Information Science, 48*(3), 236–258.

Hsieh, H. F., & Shannon, S. E. (2005). Three approaches to qualitative content analysis. *Qualitative Health Research, 15*(9), 1277–1288.

Katz, E., Blumler, J. G., & Gurevitch, M. (1973). Uses and gratifications research. *Public Opinion Quarterly, 37*(4), 509–523.

Krippendorff, K.(2004). *Content analysis: an introduction to its methodology*. Sage Publications.

Likert, R. (1932). A technique for the measurement of attitudes. *Archives of Psychology, 22*(140), 5–55.

McQuail, D. (1983). *Mass communication theory*. Sage Publications.

Moher, D., Liberati, A., Tetzlaff, J., Altman, D. G., & The PRISMA Group. (2009). Preferred reporting items for systematic reviews and meta-analyses: The PRISMA statement. *PLoS Medicine, 6*(7), Article e1000097.

Morse, J. M. (2010). Principles of mixed methods and multimethod research design. In A. Tashakkori & C. Teddle (Eds.), *Handbook of mixed methods in social and behavioral research* (2nd ed.) (pp. 189–208). Sage Publications.

Pettigrew, K. E., Fidel, R., & Bruce, H. (2001). Conceptual frameworks in information behavior. *Annual Review of Information Science and Technology, 35*, 43–78.

Poteete, A. R., Janssen, M. A., & Ostrom, E. (2010). *Working together: Collective action, the commons, and multiple methods in practice*. Princeton University Press.

Scheibe, K., Zimmer, F., & Stock, W. G. (2019). Social media usage of asylum seekers in Germany. In W. Popma & S. Francis (Eds.), *Proceedings of the 6th European Conference on Social Media* (pp. 263–272). Academic Conferences and Publishing International.

Shao, G. (2009). Understanding the appeal of user-generated media: a uses and gratification perspective. *Internet Research, 19*(1), 7–25.

Sonnenwald, D. H. (1999). Evolving perspectives of human behavior: contexts, situations, social networks and information horizons. In T. D. Wilson & D. K. Allen (Eds.), *Exploring the contexts of information behaviour* (pp. 176–190). Taylor Graham.

Sonnenwald, D. H., & Iivonen,M. (1999). An integrated human information behavior research framework for information studies. *Library and Information Science Research, 21*(4), 429–457.

Stock, W. G., & Stock, M. (2013). *Handbook of information science*. DeGruyter.

Togia, A., & Malliari, A. (2017). Research methods in library and information science. In S. Oflazoglu (Ed.), *Qualitative versus Quantitative Research* (pp. 43–64). Intech Open.

Wefugees. (n.d.). *The question & answer community for refugee topics*. Retrieved September 20, 2019, from https://www.wefugees.de/

Wilson, T. D. (1981). On user studies and information needs. *Journal of Documentation, 37*(1), 3–15.

Wilson, T. D. (1997). Information behaviour: An interdisciplinary perspective. *Information Processing & Management, 33*(4), 551–572.

Wilson, T. D. (2000). Human information behavior. *Informing Science, 3*(2), 49–56.

Zimmer, F., & Scheibe, K. (2020). Age- and gender-dependent differences of asylum seekers' information behavior and online media usage. In *Proceedings of the 53rd Hawaii International Conference on System Sciences* (pp. 2398–2407). ScholarSpace.

8 ICT and media practices for integration – A literature review

Our investigation starts with a literature review and the results of other studies will be presented. The focus of this literature review is the ICT and (online) media usage of asylum seekers in a new home country. Demonstrated are the existing body of knowledge, the observed gaps and how this study aims at being a first step in analyzing the issues at hand (Cresswell et al., 2003). The approach for the literature review was based on Fink (2019) and Moher et al. (2009). The databases used to retrieve the documents for this study are Web of Science and Scopus, as they are commonly applied for literature review and indicate a certain qualitative standard. These retrieval systems are multidisciplinary databases, meaning they cover wide areas of research. Both can be used to search for, e.g., topics, titles, abstracts, keywords, and authors. As the databases differ in the amount of journals and proceedings, it is recommended to use more than one database to enhance the overall recall (Fink, 2019). For details on the methods, see Chapter 7 (pp. 69 ff.). Based on this, the following research question (RQ) will be explored:

RQ1: How do asylum seekers use ICT, online and traditional media as well as social media for integration according to the literature?

To add the literature to the potential investigation corpora, a few requirements needed to be met. All in all, 44 documents were investigated for the content analysis of the literature review. The search, conducted on June 29, 2020, consisted of the following search queries:

Web of Science
TS = (("refugee*" OR "asylum seeker*" OR "asylee*" OR "forced migration") AND ("uses and gratifications" OR "media" OR "Internet Use" OR "ict" OR "digital" OR "communication*" OR "information behavior" OR "social media"))

Scopus
TITLE-ABS-KEY (("refugee*" OR "asylum seeker*" OR "asylee*" OR "forced migration") AND ("uses and gratifications" OR "media" OR "Internet Use" OR "ict" OR "digital" OR "communication*" OR "information behavior" OR "social media"))

https://doi.org/10.1515/9783110672022-008

8.1 Overview of data

Following, the findings are summarized to give an overview of the identified themes. In the next sections, the thematic areas are discussed in detail.

When it comes to the different types of information that are needed by asylum seekers as well as their information seeking behavior, the length of stay seems to determine those aspects (Tirosh & Schejter, 2017). Information is seen as one of the most crucial needs: "The only thing which I need is information, . . . information is like your daily bread you know, it is important" (Tirosh & Schejter, 2017, p. 10). The information types can be differentiated into "personal/survival" – information about family members in their home country and information about how to make a living; "institutional" – information about the status and rights as asylum seekers as well as other governmental or municipal service needs; "spatial/orientational" – is concerned with language, information about spatial orientation and local customs as well as information related to news from around the world. For example, regarding the institutional information needs, the regulatory changes often cause confusion among asylum seekers. "We are in a very, uhm, like undetermined situation, so we want to hear something from the government every minute, every time," as one participant states (Tirosh & Schejter, 2017, p. 11). A different example could be, as stated by another study participant, that they have been in the new country for nine months and did not know how to obtain a driver's license. Regarding the spatial challenges, even though asylum seekers acquire the country's language over time, the conversational language level is not always sufficient. For example, the bus system, in this case of Tel Aviv, was completely changed and the information was only available in Hebrew, making it impossible to understand without sufficient language skills (Tirosh & Schejter, 2017). In addition to the three identified information needs (personal/survival, institutional, and spatial/orientational) proposed by Tirosh and Schejter (2017), a study identified several other important informational areas:

> Finding appropriate work to get money to secure housing, psychological burdens suffered concerning the image of being refugees, emotional distress, lack of accessing some basic services such as education and transportation, lack of financial resources, lack of time, lack of motivation and cultural and social barriers, were significant to Syrian refugees when seeking information. (Mansour, 2018, p. 161)

Here, the information behavior is related to establishing a life in a new country as well as emotional well-being.

Based on the observed literature and studied ICT usage (see Table 8.1), different thematic areas were identified. Merz et al. (2018) observed five practices related to smartphone use by asylum seekers: 1) seeking information online, 2) communicating with family and friends abroad, 3) meeting locals, 4) meeting peers, and 5) counteracting boredom. By seeking information online, asylum seekers learn about the new country. Further, it helps with self-dependence, e.g., when acquiring information on asylum law. To communicate with family and friends abroad, smartphone applications with Internet audio and video calling capabilities and social networking sites are used. To contact and meet locals, ICT is applied often, either to initialize or maintain contacts. Further, it seems to be easier to establish new contacts online rather than offline. Additionally, meeting peers from the same home country is described as a less difficult endeavor. For entertainment, smartphones are used to watch movies and counteract boredom.

In context with digital practices, Leurs (2017) analyzed different themes. These are the 1) right to self-determination, 2) right of freedom and expression, 3) right to information, 4) right to family, and 5) right to cultural identity. Asylum seekers have the need to feel some kind of agency in a life which they cannot always control. Further, the experiences with censorship strengthens the desire for expressing oneself freely. For example, on social media, photos and videos of atrocities in Syria were removed from Facebook, Instagram, or other platforms. But, social media also helps them to connect with family across a distance. By creating public profiles on Twitter or Instagram, the diasporic attachment can be expressed, for example, by using hashtags such as "#I_love_Syria" or by sharing memories.

To overcome information challenges, the first contact point seems to be human rather than technological (Alencar & Tsagkroni, 2019; Bletscher, 2020; Nekesa Akullo & Odong, 2017; Tirosh & Schejter, 2017). Here, information is exchanged between two parties, either face to face or through online media. Furthermore, access to technology determines if someone is seen as being an opinion leader and therefore trusted more.

Next, the media mentioned in the literature as well as relevant themes were identified (Tables 8.2 and 8.3). Social media also plays an important part in giving and receiving information among asylum seekers. Mentioned social media are, e.g., Facebook, WhatsApp, and YouTube (Scheibe et al., 2019). Overall, Mokhtar and Rashid (2018) identified several needs in accordance with media usage: Media is used to fulfill the needs for entertainment, information news/updates, wealth of information, positive attitude to media, linkage, convenience, affordability, and religion. Bülbül and Haj Ismail (2019) analyzed information posted on Twitter by Syrian refugees living in Turkey. Issues found were related to obstacles while living in Turkey, such as work, dreaming of returning to the home country, and even beauty.

However, even though the Internet is used and applied by asylum seekers, some issues were observed. According to Coles-Kemp et al. (2018) as well as Mikal and Woodfield (2015), asylum seekers perceive the Internet as being dangerous. The concerns are related to frauds or being exposed to unsavory content. Therefore, the Internet is used carefully and only if needed. Further, parents often felt restricted in their ability to help and protect their children in context with mobile and digital media usage (Coles-Kemp et al., 2018). However, this can be attributed to technical literacy, as many people are not familiar with net safety tools. This in turn provides opportunities for education, offering independence and security in the application of ICT and media.

To sum up, the mentioned aspects serve as a starting point to build categories for this literature review. At some points, contradicting insights were made among the literature. However, this can be attributed to the geographical differences where the studies were conducted (see Table 8.4) or the different nationalities studied. Overall, 44 studies were consulted and examined for the categories, which are:

8.1. ICT and media usage in a new home country
8.2. Problems related to ICT and media usage
8.3. Results in accordance with the Uses and Gratifications Theory
8.4. Information exchange

Table 8.1: Identified ICT in the literature.

ICT	Literature
Computer	(Alam & Imran, 2015; Almohamed & Vyas, 2019; Bacishoga et al., 2016; Bletscher, 2020; Díaz Andrade & Doolin, 2019; Mansour, 2018; Oduntan & Ruthven, 2017; Sabie & Ahmed, 2019; Scheibe et al., 2019; Simko et al., 2018; Tirosh & Schejter, 2017)
Internet	(AbuJarour et al., 2018; Ahmad, 2020; Bletscher, 2020; Dhoest, 2020; Köster et al., 2018; Leurs, 2017; Merisalo & Jauhiainen, 2020; Oduntan & Ruthven, 2017; Simko et al., 2018; Stiller & Trkulja, 2018; Tirosh & Schejter, 2017)
Landline	(Mikal & Woodfield, 2015; Scheibe et al., 2019; Shrestha-Ranjit et al., 2020; Tudsri & Hebbani, 2015; Yun et al., 2016)
Laptop	(Alam & Imran, 2015; Kaufmann, 2018; Leurs, 2017; Scheibe et al., 2019; Stiller & Trkulja, 2018; Tirosh & Schejter, 2017; Witteborn, 2019)
Mobile phone, smartphone	(AbuJarour & Krasnova, 2017; Bacishoga et al., 2016; Kaufmann, 2018; Kneer et al., 2019; Koh et al., 2018; Mansour, 2018; Stiller & Trkulja, 2018; Tirosh & Schejter, 2017; Tudsri & Hebbani, 2015; Witteborn, 2019)

Table 8.1 (continued)

ICT	Literature
Others: Internet café, flash drive	(Tirosh & Schejter, 2017)
Radio	(Mansour, 2018; Scheibe et al., 2019; Tudsri & Hebbani, 2015)
Smartphone	(AbuJarour & Krasnova, 2017; AbuJarour & Krasnova, 2018; Ahmad, 2020; Alam & Imran, 2015; Alencar & Tsagkroni, 2019; Almohamed & Vyas, 2019; Bletscher, 2020; Dhoest, 2020; Graf, 2018; Kaufmann, 2018; Kneer et al., 2019; Kutscher & Kreß, 2018; Leurs, 2017; Marlowe, 2019; McCaffrey & Taha, 2019; Merisalo & Jauhiainen, 2020; Merz et al., 2018; Mikal & Woodfield, 2015; Sabie & Ahmed, 2019; Scheibe et al., 2019; Simko et al., 2018; Witteborn, 2019)
Tablet	(Graf, 2018; Kaufmann, 2018; McCaffrey & Taha, 2019; Sabie & Ahmed, 2019; Scheibe et al., 2019)
Television	(Díaz Andrade & Doolin, 2019; Kaufmann, 2018; Mansour, 2018; Mokhtar & Rashid, 2018; Scheibe et al., 2019; Tirosh & Schejter, 2017; Tudsri & Hebbani, 2015)

Table 8.2: Identified media in the literature.

Traditional and online media	Literature
Books, magazine	(Dhoest, 2020; Mansour, 2018; Scheibe et al., 2019)
Dating app (e.g., Badoo, hi5, Meetone, Twoo, Grindr)	(AbuJarour & Krasnova, 2018; Dhoest, 2020; Merz et al., 2018; Witteborn, 2019)
Education (e.g., video tutorials, moodle platform, app)	(AbuJarour & Krasnova, 2018; Merz et al., 2018)
E-mail	(Alam & Imran, 2015; Alencar & Tsagkroni, 2019; Bacishoga et al., 2016; Díaz Andrade & Doolin, 2019; Mikal & Woodfield, 2015; Mokhtar & Rashid, 2018; Sabie & Ahmed, 2019; Scheibe et al., 2019; Simko et al., 2018; Stiller & Trkulja, 2018; Tirosh & Schejter, 2017; Udwan et al., 2020)
E-payment app	(AbuJarour et al., 2019; Coles-Kemp et al., 2018; Sabie & Ahmed, 2019)

Table 8.2 (continued)

Traditional and online media	Literature
Gaming	(Ahmad, 2020; Merz et al., 2018; Mikal & Woodfield, 2015; Scheibe et al., 2019)
GPS, navigation app (e.g., DB navigator app, location finding app, Google maps)	(AbuJarour et al., 2019; Ahmad, 2020; Alencar & Tsagkroni, 2019; Coles-Kemp et al., 2018; Díaz Andrade & Doolin, 2019; Duarte et al., 2018; Kaufmann, 2018; Kutscher & Kreß, 2018; McCaffrey & Taha, 2019; Merz et al., 2018; Sabie & Ahmed, 2019; Scheibe et al., 2019)
Internet radio	(Díaz Andrade & Doolin, 2019; Merz et al., 2018; Tirosh & Schejter, 2017)
Mobile instant messenger (e.g., Yahoo! Messenger, WickR)	(Bacishoga et al., 2016; Leurs, 2017; Tirosh & Schejter, 2017; Tudsri & Hebbani, 2015)
News website, (online) news channel	(Díaz Andrade & Doolin, 2019; Mokhtar & Rashid, 2018; Scheibe et al., 2019; Stiller & Trkulja, 2018; Tirosh & Schejter, 2017; Tudsri & Hebbani, 2015)
Newspaper	(Mansour, 2018; Marlowe, 2019; Scheibe et al., 2019; Tirosh & Schejter, 2017; Tudsri & Hebbani, 2015)
Official websites (German government website, UNHCR)	(Köster et al., 2018; Mokhtar & Rashid, 2018)
Other: Ethnic media, online webpage, various apps (e.g., life counseling app, prayer times app, design software, ebay), Wikipedia, SMS	(AbuJarour et al., 2019; AbuJarour & Krasnova, 2017; Coles-Kemp et al., 2018; Graf, 2018; Mokhtar & Rashid, 2018; Sabie & Ahmed, 2019; Scheibe et al., 2019)
Search engine (e.g., Google search engine)	(Alencar, 2018; Díaz Andrade & Doolin, 2019; Köster et al., 2018; Scheibe et al., 2019; Stiller & Trkulja, 2018; Tirosh & Schejter, 2017)
Translation services (e.g., app, website, Google translate, Microsoft translator)	(AbuJarour et al., 2018; AbuJarour et al., 2019; AbuJarour & Krasnova, 2017; AbuJarour & Krasnova, 2018; Alencar & Tsagkroni, 2019; Almohamed & Vyas, 2019; Coles-Kemp et al., 2018; Duarte et al., 2018; Köster et al., 2018; McCaffrey & Taha, 2019; Merz et al., 2018; Scheibe et al., 2019; Tirosh & Schejter, 2017)

Table 8.3: Identified social media in the literature.

Social media	Literature
Facebook	(AbuJarour et al., 2019; AbuJarour & Krasnova, 2017; AbuJarour & Krasnova, 2018; Ahmad, 2020; Alencar, 2018; Alencar & Tsagkroni, 2019; Almohamed & Vyas, 2019; Bletscher, 2020; Dhoest, 2020; Graf, 2018; Kaufmann, 2018; Kneer et al., 2019; Köster et al., 2018; Kutscher & Kreß, 2018; Leurs, 2017; Mansour, 2018; Marlowe, 2019; Marlowe, 2020; McCaffrey & Taha, 2019; Merz et al., 2018; Mikal & Woodfield, 2015; Mokhtar & Rashid, 2018; Scheibe et al., 2019; Simko et al., 2018; Stiller & Trkulja, 2018; Tirosh & Schejter, 2017; Tudsri & Hebbani, 2015; Udwan et al., 2020; Witteborn, 2019)
Instagram	(AbuJarour et al., 2019; Ahmad, 2020; Alencar, 2018; Kneer et al., 2019; Leurs, 2017; Marlowe, 2019; McCaffrey & Taha, 2019; Mokhtar & Rashid, 2018; Scheibe et al., 2019)
Professional (e.g., LinkedIn, Xing)	(Alencar, 2018; Alencar & Tsagkroni, 2019; Stiller & Trkulja, 2018)
Other: live streaming services, reddit, TikTok, 9GAG	(Scheibe et al., 2019)
Skype	(Ahmad, 2020; Dhoest, 2020; Díaz Andrade & Doolin, 2019; Kaufmann, 2018; Kutscher & Kreß, 2018; Merz et al., 2018)
Snapchat	(AbuJarour et al., 2019; Ahmad, 2020; Kneer et al., 2019; Leurs, 2017; Scheibe et al., 2019; Udwan et al., 2020)
Twitter	(Ahmad, 2020; Alencar, 2018; Bülbül & Haj Ismail, 2019; Marlowe, 2019; Scheibe et al., 2019; Stiller & Trkulja, 2018)
Viber	(Ahmad, 2020; Alencar, 2018; Almohamed & Vyas, 2019; Bletscher, 2020; Kaufmann, 2018; Kutscher & Kreß, 2018; Marlowe, 2019; Merz et al., 2018; Simko et al., 2018; Witteborn, 2019)
WhatsApp	(AbuJarour et al., 2019; AbuJarour & Krasnova, 2017; AbuJarour & Krasnova, 2018; Ahmad, 2020; Alencar, 2018; Alencar & Tsagkroni, 2019; Almohamed & Vyas, 2019; Bletscher, 2020; Coles-Kemp et al., 2018; Kaufmann, 2018; Kneer et al., 2019; Kutscher & Kreß, 2018; Leurs, 2017; Marlowe, 2020; McCaffrey & Taha, 2019; Merz et al., 2018; Mokhtar & Rashid, 2018; Sabie & Ahmed, 2019; Scheibe et al., 2019; Simko et al., 2018; Witteborn, 2019)

Table 8.3 (continued)

Social media	Literature
Other messengers (e.g., Facebook Messenger, Tango, IMO, Telegram, Line, WeChat)	(AbuJarour et al., 2019; Ahmad, 2020; Alencar, 2018; Kaufmann, 2018; Marlowe, 2019; McCaffrey & Taha, 2019; Merz et al., 2018; Mokhtar & Rashid, 2018; Scheibe et al., 2019; Witteborn, 2019)
YouTube	(AbuJarour et al., 2019; AbuJarour & Krasnova, 2018; Ahmad, 2020; Merz et al., 2018; Mokhtar & Rashid, 2018; Simko et al., 2018; Udwan et al., 2020)

Table 8.4: Countries of included studies.

Country of study	Literature
Australia	(Alam & Imran, 2015; Almohamed & Vyas, 2019; Koh et al., 2018; Tudsri & Hebbani, 2015; Walker et al., 2015)
Austria	(Kaufmann, 2018; Merz et al., 2018)
Belgium	(Dhoest, 2020)
Egypt	(Mansour, 2018)
Germany	(AbuJarour et al., 2018; AbuJarour et al., 2019; AbuJarour & Krasnova, 2017; AbuJarour & Krasnova, 2018; Duarte et al., 2018; Graf, 2018; Köster et al., 2018; Kutscher & Kreß, 2018; Scheibe et al., 2019)
Greece	(Merisalo & Jauhiainen, 2020)
Hong Kong	(Witteborn, 2019)
Iran	(Merisalo & Jauhiainen, 2020)
Israel	(Tirosh & Schejter, 2017)
Italy	(Merisalo & Jauhiainen, 2020)
Jordan	(Merisalo & Jauhiainen, 2020)
Canada	(Sabie & Ahmed, 2019)
Lebanon	(Ahmad, 2020)
Malaysia	(Mokhtar & Rashid, 2018)

Table 8.4 (continued)

Country of study	Literature
Netherlands	(Alencar & Tsagkroni, 2019; Kneer et al., 2019; Leurs, 2017; Udwan et al., 2020)
New Zealand	(Shrestha-Ranjit et al., 2020)
South Africa	(Bacishoga et al., 2016)
Sweden	(Coles-Kemp et al., 2018; Graf, 2018)
Turkey	(Bülbül & Haj Ismail, 2019; Merisalo & Jauhiainen, 2020)
U.S.	(Bletscher, 2020; Mikal & Woodfield, 2015; Simko et al., 2018; Yun et al., 2016)
Uganda	(Nekesa Akullo & Odong, 2017)
United Kingdom	(Oduntan & Ruthven, 2017)

8.2 ICT and media usage in a new home country

When establishing oneself in a new country, finding employment is challenging, therefore, buying different technology appliances, including Internet access devices, becomes difficult (Alam & Imran, 2015). As a result, accessing important information can be problematic (Nekesa Akullo & Odong, 2017). Tirosh and Schejter (2017) state how asylum seekers sometimes sit in restaurants with channels from their home country being broadcasted on TVs. Certain stores sell phone cards, music discs, and DVDs from the asylum seekers' home country, the owner helping many of the customers who do not know how to access the Internet. Interviewees also stated that they do not own a computer and therefore relied on Internet cafés. Others even use 3G mobile services to access the Internet on their laptops. If no cable modems or routers are accessible, Wi-Fi hotspots serve as alternatives (Tirosh & Schejter, 2017). A study exploring challenges of asylum seekers accessing the heavily technological infrastructure in Canada arrived at interesting results (Sabie & Ahmed, 2019). All asylum seekers owned a smartphone and had a data plan, except for one participant. Even the children, one being only eight years old, owned a smart device. Mobile phones are used on a regular basis and, surprisingly, other digital devices are seldom utilized. But, all interviewees stated that they had access to Wi-Fi and at least one computer and or tablet at home, whereas only four of the interview participants used a computer. Similarly, the library and community centers offer free Wi-Fi and computer access. Being able to connect to the Internet is simultaneously seen as being connected to the world (Coles-Kemp et al., 2018). The mobile phone is present in all areas of life and often the first thing the participants look at in the morning (Coles-Kemp et al., 2018).

A closer look will be taken at different age groups and gender-dependent differences and similarities regarding ICT application when establishing oneself in a new home country. Regarding age differences, a study conducted by Merisalo and Jauhiainen (2020) concerning the Internet use and smartphone ownership of different age groups concludes that the older asylum seekers are, the less likely they are to use the Internet. Another study confirms this distribution, the younger generation has more ICT related skills:

> For us it's a lot easier, but for our parent's, that's a barrier . . . if I explain to my mum that I'm going on the Internet and I'm trying to find this information, she like gets out the Yellow Pages and calls the doctor. I'm like no I can go and type it in and look for a doctor and look for their schedule, and she finds that hard to comprehend. So it's just . . . getting on the Internet and finding certain things is a problem. (Alam & Imran, 2015, p. 356)

Further, children at school age, teenagers and early-twenties to young adults, have a higher access rate to technology than even their older siblings and consequently parents and grandparents (Bletscher, 2020). When focusing on younger participants and their usage of social networking sites and messaging services, WhatsApp to communicate with others and Facebook to seek and share information are used the most often. Twitter, Instagram, and Snapchat, however, are rarely used. YouTube, on the other hand, is applied a lot for entertainment and leisure purposes.

Also, there seems to be no gender-based digital divide in regards to Internet users (Merisalo & Jauhiainen, 2020). In contrast, Ahmad (2020) reports that, according to their study of asylum seekers' social media and device usage, of the 40% who do not own a device, the majority of those are young unmarried females between 15 and 18 years. Alas, the Internet was not the focus, but smartphone ownership. This group uses smartphones owned by their parents or siblings. It came to light that, often, the male person of the household is responsible for purchasing the phone for the children and wife. Furthermore, once a sibling or parent upgrades to a newer model, the older phone is given to the younger family member. This notion is supported by a study of smartphone use of asylum seekers in New Jersey (McCaffrey & Taha, 2019): Men tend to have the newest models of phones. Whereas they and older school-aged children also own data plans, many younger children do have their own device, but no data plan. Some women had two phones, a smartphone without a data plan and a welfare phone, which is a flip phone provided by charity programs. When the husband left for work or similar duties, he took the data plan device with him.

The app distribution among refugees in the new country shows interesting insights. According to AbuJarour et al. (2019), a study about the most common apps on the mobile phone home screens reveals that the most often user-installed app is WhatsApp. Other popular apps are Facebook Messenger, YouTube, Facebook, and Instagram. This was compared to a German user group and the results indicate that asylum seekers install Facebook and YouTube significantly more often than the German participants. Moreover, the most helpful apps seem to be social media apps, e-payment apps, location finding apps, translation apps, and apps for supporting the completion of school work (Coles-Kemp et al., 2018).

8.3 Problems related to ICT and media usage

A few problems in relation to ICT and media usage were mentioned in the literature. According to Sabie and Ahmed (2019), all of their participants were not familiar with most applications they had to use on a regular basis in their new home country. Examples were e-mails or banking. For some, they did not know about these applications or did not use them before. This is also reflected in their information behavior: doing things online or digitally and looking for information on the Internet is not familiar. Often, there is no awareness on how or where to look for information. This is echoed by observations made by Mikal and Woodfield (2015) as well as Tirosh and Schejter (2017). They found that the language skills and digital literacy which are necessary to locate relevant information is not something that everyone has – it cannot be taken for granted. When asked about if one participant could find, for example, the business hours of a certain institution online, they would not be able to do so (Mikal & Woodfield, 2015). When it comes to accessing and finding information, even if formal networks such as local municipalities or NGOs offer digital initiatives regarding information, in this case, the labor market, asylum seekers did not know about these offerings (Alencar & Tsagkroni, 2019), showcasing how knowing where to access certain information is crucial.

This problem ties in with the accessibility of ICTs. Even though ICTs offer a multitude of advantages for successful integration, it is argued that these are not widely available for asylum seekers (Bletscher, 2020). Further, fluency of a foreign language, most often English, impacts the use of information technologies. Surprisingly, this also has another side. Because asylum seekers are able to apply ICT and connect to other similar ethnic groups, this has discouraging effects on developing English fluency. However, following Díaz Andrade and Doolin (2019), participants were highly aware about their ICT skills, and were therefore more employable and business savvy. Those who do not use the Internet due to a lack of availability knew they were lacking opportunities: "It is very important as all the assignments are through the Internet and without the Internet you can't get your work done. For instance, we can do all our transactions on the Internet such as paying bills, banking and shopping. It saves a lot of time" (Alam & Imran, 2015, p. 354). Further, the Internet provides asylum seekers with a sense of independence. Through applying those media and ICTs, they become more confident and have more freedom (Bletscher, 2020). Merz et al. (2018) state how translation apps and online dictionaries are supportive in being self-dependent. For example, when encountering situations in which one needs to speak the new country's language, these services can be used to communicate efficiently. Additionally, it was mentioned how a few asylum seekers do not trust others when it comes to information and therefore use ICTs (Bletscher, 2020). When it comes to who to trust about information, there is conflicting literature. Merz et al. (2018) state that refugees use online sources to verify the information obtained by word-of-mouth. However, concerning the aspect of independence, some participants admit how they rely too much on resettlement agencies to use ICTs for them, limiting self-sufficiency in technology use (Bletscher, 2020). This results in codependency and reliance on these agencies. It is known that those asylum seekers who have access to other information sources because of their language skills and a broader social network in their new home country, rely less on social media to find information, for example, regarding legal or organizational information (Köster et al., 2018). Even though this phenomenon could be observed, the use of social media still contributes to social inclusion by providing information and therefore helping with integration (Köster et al., 2018).

The adoption of media and technology leads to some problems (Simko et al., 2018). Since using a certain new technology to accomplish a goal, like writing an e-mail or learning a new language, using it securely and privately is sometimes not a top priority. This is made even more difficult by the fact that many asylum seekers do not have a computer at home (Simko et al., 2018). Further, Mikal and Woodfield (2015) as well as Coles-Kemp et al. (2018) observed how their respondents perceived the Internet as being dangerous, halting them from using it. However, with the right digital literacy tools these perceived dangers can be overcome (Mikal & Woodfield, 2015).

8.4 Results in accordance with the uses and gratifications theory

Upon reviewing the literature, the studies were categorized according to the Uses and Gratifications Theory (U>) after Katz et al. (1973), McQuail (1983) as well as Shao (2009). As the nature of this book is the analysis of the motives to use ICT as well as media and social media, this study aims at finding the motives of asylum seekers to apply such ICT and media. As the U> is widely established, it is fitting to serve such a purpose. Furthermore, it was distilled what information these vulnerable groups are most interested in when trying to establish a new life. The applied information behavior, such as, what kind of ICT or media was utilized to serve those needs, was analyzed as well.

Information

In the following, the literature concerning themes in relation to information will be presented. These themes were further divided in subcategories for an easier overview. The identified themes are: language learning; education and employment; health; law; news; housing; everyday information and tasks.

Language learning. A few papers could be identified dealing with language learning in relation to media and ICT (AbuJarour & Krasnova, 2017, 2018; Alencar, 2018; Almohamed & Vyas, 2019; Díaz Andrade & Doolin, 2019; Koh et al., 2018; McCaffrey & Taha, 2019; Merz et al., 2018; Mikal & Woodfield, 2015; Nekesa Akullo & Odong, 2017; Sabie & Ahmed, 2019; Tudsri & Hebbani, 2015; Witteborn, 2019). According to AbuJarour and Krasnova (2018), Alencar (2018) as well as Merz et al. (2018), learning the new country's foreign language is the most important concern for asylum seekers. Learning a new language correlates with the desire to understand a new culture (Alencar, 2018; Merz et al., 2018). According to Alencar (2018), participants reported on how they applied technologies for language and cultural learning even before they received their residence permit. This can also be due to the unfortunate fact that, depending on the new country, asylum seekers are not allowed to participate in education programs, including language courses, until the asylum process is completed (AbuJarour & Krasnova, 2017). Therefore, it is not surprising that asylum seekers turn to ICTs to acquire different educational goals. Of course, where applicable, visiting language schools are important for language learning (AbuJarour & Krasnova, 2018; Mikal & Woodfield, 2015). But, additional means are needed, ranging from several media to online offers. When it comes to traditional media, Almohamed and Vyas (2019) describe how participants use media such as dictionaries to learn the definition of a word and even a board to write new words to remember them better. Asylum seekers seem to rely heavily on ICT to learn the new country's language (AbuJarour & Krasnova, 2017; Almohamed & Vyas, 2019). For example, Díaz Andrade and Doolin (2019) describe how the computer at home is used for daily language lessons. Furthermore, it is possible to contact teachers via e-mail this way. Even a phone and laptop can be utilized for language acquisition as described by Witteborn (2019). Another favored method to acquire a new language is the application of several online media (AbuJarour & Krasnova, 2018; Almohamed & Vyas, 2019; Merz et al., 2018; Mikal & Woodfield, 2015). Merz et al. (2018) describe how refugees search for online materials in order to help with learning German by self-study. For example, educational websites that were provided by the instructors of language courses were acknowledged (Mikal & Woodfield, 2015). Furthermore, dedicated websites concerned with language learning were mentioned by Merz et al. (2018), as well as specialized apps (AbuJarour & Krasnova, 2017, 2018), and online dictionaries:

> I used my computer to improve my English through using different types of dictionaries. In the past, I used hard copy dictionaries, which are not easy to find the meaning of words, but nowadays I have different types of digital dictionaries, so I can use three to four different types of dictionary on the computer. Sometimes I use the Internet to find specific meanings for the words or for sentences; and to translate the text you may need more than one dictionary. (Almohamed & Vyas, 2019, p. 41)

AbuJarour and Krasnova (2017) also mention some apps which help asylum seekers to connect with volunteers of a new country to learn the native language, underlining the importance of local contacts. Furthermore, refugees seem to access several e-learning channels to learn a new language (AbuJarour & Krasnova, 2018). Here, YouTube was mentioned several times (AbuJarour & Krasnova, 2017, 2018; Alencar, 2018). "Social media is helping a lot to learn the Dutch, because you can search for You-Tube, and follow so many videos, and you can start your first step of learning the Dutch language" (Alencar, 2018, p. 1598). AbuJarour and Krasnova (2017) describe how most of their interviewees mentioned a YouTube channel hosted by an Arabic teacher who posts lessons in Arabic to teach German. Other social media seem to be a favored method for language acquisition (Alencar, 2018). AbuJarour and Krasnova (2018) acknowledge applications such as Facebook and WhatsApp for this; Merz et al. (2018) also mention video tutorials on Facebook. When asked about social media, one participant stated: "I do not know if it helps us for adaptation, but we can learn the language of the new country" (Alencar, 2018, p. 1597). Further, it was mentioned how the creation of Facebook pages and groups that include information on culture, language and traditions of every culture could be really helpful to asylum seekers (Alencar, 2018). Of course, native media of the new country play an important part as well. For example, Merz et al. (2018) state how refugees listen to Austrian radio broadcasts to not even learn the country's native language, German, but also familiarize themselves with the Austrian dialect. Tudsri and Hebbani (2015) observed that those who have limited language proficiency resorted to watching, in this case, English language media. Those who have no interest in improving their skills resort to media in their own native language. When it comes to children and their language acquisition, a study observed how the younger ones learn words for colors with the help of an iPad game (McCaffrey & Taha, 2019).

Education and employment. Coming to education and employment, a few observations were made (AbuJarour & Krasnova, 2018; Ahmad, 2020; Bacishoga et al., 2016; Díaz Andrade & Doolin, 2019; Koh et al., 2018; Mansour, 2018; Merisalo & Jauhiainen, 2020; Mikal & Woodfield, 2015; Nekesa Akullo & Odong, 2017; Sabie & Ahmed, 2019; Stiller & Trkulja, 2018; Tirosh & Schejter, 2017). According to Nekesa Akullo and Odong (2017), 35% of their study participants need information on education and jobs to improve their well-being. In this context, when seeking information, finding appropriate work to earn money and housing are significant to Syrian asylum seekers (Mansour, 2018). According to AbuJarour and Krasnova (2018), one participant states: "[T]he most important thing is deciding to join a Master's program at Berlin University of Technology using my smartphone. I visited their website and I found a special program for refugees where they can join lectures as guests" (p. 7). Again, this highlights the importance of ICT in a new environment. In context with education acquisition, some countries make it difficult to get previous earned credentials approved in the home country, requiring asylum seekers to gain the degree again in the new country (Sabie & Ahmed, 2019). An Internet connection can help in acquiring education, as was explained by a Sudanese asylum seeker (Tirosh & Schejter, 2017). The Internet was also mentioned when it comes to finding both volunteer and paid employment in the United States (Mikal & Woodfield, 2015). Furthermore, the Internet is very helpful in monitoring the job market and which skills and documents are required to apply for a particular type of job as was explained by Díaz Andrade and Doolin (2019) on a study of asylum seekers living in New Zealand. Additionally, the mobile phone was mentioned in context with employment. According to a study by Bacishoga et al. (2016), the mobile phone is stressed as being very important in order to be contactable for employment purposes. This is especially true for asylum seekers who are self-employed. The mobile phone is used to recruit customers and welcome them to be future customers. Even a study on gender- and age-dependent aspects concerned with job opportunities was observed. According to Merisalo and Jauhiainen (2020), male respondents are 1.5 times more likely than female respondents to seek information online about work opportunities. This statement is true for younger asylum seekers (19 to 29 years old) as well – they are more likely to search for work opportunities online than middle aged or older asylum seekers. All in all, refugees are aware of how ICT skills help to make oneself more employable or are even useful to set up a business (Díaz Andrade & Doolin, 2019). However, a self-reported skill assessment on web browser, Internet search engine, and social network usage showcased that the participants rated their skills to be above average (Stiller & Trkulja, 2018). But, after completing some tasks related to digital literacy skills during an observatory study, the researchers found that the perceived skills were overestimated as the participants demonstrated lower skills in practice. Observed issues were related to information seeking processes such as formulating advanced search queries which were often not known about or orienting oneself on a website.

Health. Literature on health information behavior was identified (Almohamed & Vyas, 2019; Graf, 2018; Mikal & Woodfield, 2015; Nekesa Akullo & Odong, 2017; Scheibe et al., 2019; Tirosh & Schejter, 2017; Udwan et al., 2020). Concerning health information, according to Nekesa Akullo and Odong (2017), 95% of their study participants reported that they need information on health services. For example, mothers wanted to know how they can get access to health services for their children and themselves. Scheibe et al. (2019) arrive at similar findings. Here, the topic health dominates all other information categories in its importance. When it comes to ICT and media, Google Translate was mentioned as very helpful when visiting doctor appointments (Almohamed & Vyas, 2019). Furthermore, the Internet is helpful in accessing information on health in general (Tirosh & Schejter, 2017) or even certain medications (Mikal & Woodfield, 2015). According to a study on social media for social support, health, and identity, Udwan et al. (2020) state how the first step in seeking health information is to look up doctors' contact details and nearby hospitals or clinics via Internet-based platforms. "We can contact the GP via email, or we can make an appointment online without any difficulty and would save time and make life easy" (Udwan et al., 2020, p. 7). One feature that seems to be particularly convenient is the geolocalization option which is added to most social media platforms and mobile applications. They are helpful to find and evaluate the closest medical clinic (Udwan et al., 2020). In this context, social media platforms were mentioned as well. Here, asylum seekers have the opportunity to share their experiences regarding health procedures and give assistance in the new country. A great example would be the Facebook community group "الطبية في هولندا medical in Nederland" which is focused on the needs of the Syrian community in the Netherlands with 16,000 followers. "I ask questions on the Facebook page 'medical in Nederland' about health issues, when I get sick, I can ask about medicines, different diseases, comparing health insurance, death insurance, healthy food, useful information about pregnancy, for children, etc." (Udwan et al., 2020, p. 7). If credibility is concerned, the information is preferably discussed via social platforms with trusted social networks or family members with medical backgrounds. In this context, content on social media is not only consumed by asylum seekers, but also produced. For example, one asylum seeker, working as a pharmacist, created a page on Facebook and sees it as his duty to help his people. The same person also produces YouTube videos about those topics. Albeit not mentioned often in the literature, mental health was brought up by Graf (2018). Here, one participant shared that he uses a life counseling app to manage his mental health. For him, he could depend on the app to give him valuable advice and support.

Law. Another type of information which is important to asylum seekers concerns the legal system of a new home country (Alencar & Tsagkroni, 2019; Duarte et al., 2018; Köster et al., 2018), especially the asylum status (AbuJarour & Krasnova, 2017; Mansour, 2018; Merz et al., 2018; Mokhtar & Rashid, 2018; Witteborn, 2019). Here, ICTs and media are used to retrieve information related to these topics. However, finding and understanding the content of those laws can be troublesome. According to Köster et al. (2018), their interviewees state how difficult it can be to assess this kind of information. Accordingly, German government sites, for example, are not always provided in the asylum seekers' native language and therefore require advanced language skills: "At first, I try to visit the German government websites, and if it is difficult I try to ask a German person and if I can't find the answer I'm looking for, I try to ask in Facebook groups; they can sometimes help" (Köster et al., 2018, p. 5). Therefore, following the statements of the study's participants, asking other natives, e.g. Germans, for help, is considered to be beneficial in this situation. However, as is also evident, asking native speakers is not always possible. Here, social media serves as an alternative information source. In this case, asylum seekers would refer to Facebook pages particularly specialized for Syrians and their information needs. Another mentioned Facebook page is WDRforyou, a member of the consortium of German public-broadcasting institutions. Similar findings regarding social media were made by Alencar and Tsagkroni (2019). Here, the connections that were made on social media are helpful in sharing information with asylum seekers to avoid breaking specific rules and laws in the Netherlands. Asylum seekers rely on their smartphones to access and translate official government websites (AbuJarour &

Krasnova, 2017). Accordingly, asylum regulations change frequently and are published only in the local language. Asylum seekers need to follow a lot of administrative steps involving different governmental offices to finalize the asylum process, e.g., getting a residence permit or health insurance. When obtaining information on the current refugee status, some asylum seekers access the official UN website. "For information about refugee status, we check the website of UNHCR regularly or email them, but they are very slow to reply" (Mokhtar & Rashid, 2018, p. 81). Here, in contrast to findings by Alencar and Tsagkroni (2019) as well as Köster et al. (2018) on general information about the law of a new country, this information cannot be obtained from social media. However, an asylum seeker living in Malaysia proposed the idea of a social media page where this information can be shared: "I think Facebook. Like we have applied for asylum for several years now, there should be a place where we can get an update of our case there. Maybe some website, we can have a link to it. Information could be posted on it, or e-mailed" (Mokhtar & Rashid, 2018, p. 77). This statement also highlights that it is not always obvious where to find the needed information, in this case, updates on the asylum status.

News. Literature on news was also identified (AbuJarour & Krasnova, 2017; Alencar & Tsagkroni, 2019; Dhoest, 2020; Díaz Andrade & Doolin, 2019; Graf, 2018; Kaufmann, 2018; Mansour, 2018; Marlowe, 2019; Merisalo & Jauhiainen, 2020; Mikal & Woodfield, 2015; Mokhtar & Rashid, 2018; Scheibe et al., 2019; Tirosh & Schejter, 2017; Tudsri & Hebbani, 2015). When asked about what kind of news the participants searched for, it was mainly related to news and other information from their home country (Mansour, 2018; Merisalo & Jauhiainen, 2020; Mikal & Woodfield, 2015; Scheibe et al., 2019). According to Scheibe et al. (2019), many of the study participants mentioned an interest in information concerning their home country (13 out of 17), but even more in information concerning the new country (15 out of 17). Scheibe et al. (2019) observed that news was read in the study participants' native language (13 from 17 participants) but also in the new home country's native language (9 from 17 participants). Only one participant was interested in English news. This limitation on the English language was observed by Díaz Andrade and Doolin (2019) as well. Graf (2018) arrived at quite different results. In their case, news programs in the native language of the new country, here, Swedish or German news channels, are not used because of language barriers. Whereas back in the home country, the radio, newspapers, and television were available in the asylum seekers' native language, accessing those kinds of media in a new country poses a challenge. Therefore, ICT mediated news consumption, for example via the Internet, presents an alternative to overcome language barriers (Dhoest, 2020; Díaz Andrade & Doolin, 2019; Mokhtar & Rashid,

2018; Tirosh & Schejter, 2017; Tudsri & Hebbani, 2015). Mobile phones or phone calls (AbuJarour & Krasnova, 2017; Díaz Andrade & Doolin, 2019; Tudsri & Hebbani, 2015) and the computer (Díaz Andrade & Doolin, 2019) are additionally used for news consumption in conjunction with the Internet. "As a refugee, I use my smartphone to learn, to stay in touch with my family, to navigate from one place to another, and to catch up with recent changes and news" (AbuJarour & Krasnova, 2017, p. 1796). As one participant states: "We can know the news from around the world . . . I watch news every morning on the computer" (Díaz Andrade & Doolin, 2019, p. 155). Other further mentioned applications to access news are the TV (Alencar & Tsagkroni, 2019; Díaz Andrade & Doolin, 2019; Tirosh & Schejter, 2017; Tudsri & Hebbani, 2015), online or offline radio (Díaz Andrade & Doolin, 2019; Mansour, 2018; Tirosh & Schejter, 2017), and news websites (Díaz Andrade & Doolin, 2019; Tirosh & Schejter, 2017). In the case of Mansour (2018), one-third of the Syrian study participants use both TV and radio programs to receive information, especially on the current status of the Syrian crisis. Preferred were programs from a political and social perspective. Concerning traditional print media, a few papers were mentioned (Mansour, 2018; Tirosh & Schejter, 2017; Tudsri & Hebbani, 2015). Here, magazines and books (Mansour, 2018), and newspapers (Mansour, 2018; Tirosh & Schejter, 2017) were utilized. However, according to Tudsri and Hebbani (2015), even though newspapers and the radio are used to obtain information, they are not popular media choices. Only two of their 29 study participants reported on using the radio to listen to news. Focusing on specific news channels and programs, a few studies were identified: Asylum seekers consume local media and watch local television news programs, for example, a study participant who had gained a working knowledge of Hebrew reported watching Channel 2 news in Israel (Tirosh & Schejter, 2017). In this study's case, the asylum seekers also purchased and read the local news post, The Jerusalem Post. Local news origins, either being online or offline media, were reported as being sources of information (Díaz Andrade & Doolin, 2019; Tudsri & Hebbani, 2015). For example, Tudsri and Hebbani (2015) state how a few study participants like to watch Australian news programs to learn more about their new home country Australia. According to Díaz Andrade and Doolin (2019, p. 155), one study participant, now living in Nepal, states: "[L]isten to . . . a Nepalese radio every Saturday. It is a radio of what is happening in the world. Other mentioned news sources are South Sudanese newspapers on the Internet and private Sudanese satellite channel TV shows (Tirosh & Schejter, 2017). This also applies to news sources in English. According to Tirosh and Schejter (2017) news sources include local newspaper websites in English. Further, a Nigerian asylum seeker reported watching the following array of English-language programs and channels: The IBA English news program, Middle East Television, CNN, and

Fox News. In one particular case, even though the study participant is able to understand English speaking news, he additionally relies on ICT to access supplementary online information to enhance his understanding of the news items seen on TV (Díaz Andrade & Doolin, 2019). "If [the topic] is interesting, we can type [it] on the Internet . . . For example, in Al-Jazeera News, they say 'If you need more information, go to the website'" (Díaz Andrade & Doolin, 2019, p. 155). But also, as mentioned earlier, where possible, participants like to access news sources in their native language (Díaz Andrade & Doolin, 2019; Tirosh & Schejter, 2017; Tudsri & Hebbani, 2015). Examples mentioned are BBC online (in Dari, Persian, and Urdu), SBS World news, Tolo news (an online Afghan news site), and Geo News (Pakistani online news website) in Urdu and English (Tirosh & Schejter, 2017). According to Díaz Andrade and Doolin (2019), BBC News is also mentioned, this time in Burmese. Other mentioned news sources are Radio Assenna from Eritrea (Tirosh & Schejter, 2017). However, some participants state they are aware of the circulation of fake news distributed by media outlets overseen by the government (Mokhtar & Rashid, 2018). "We can't believe these media, but I still check them to know what bogus news they are broadcasting there. I keep myself alerted to what kind of fake news they are spreading" (Mokhtar & Rashid, 2018, p. 83). To overcome this challenge, some like to use social media and messengers like Facebook and WhatsApp to compare received information with friends and family.

Even though these news sources, such as BBC, Al Jazeera, CNN, and Fox News, are trusted, many study participants state that the news reached them quicker through their social media pages (Marlowe, 2019). The social media application Facebook seems to be especially popular for this (AbuJarour & Krasnova, 2017; Graf, 2018; Kaufmann, 2018; Marlowe, 2019; Mokhtar & Rashid, 2018), as well as Twitter (Marlowe, 2019). Facebook offers profiles of news outlets, helping asylum seekers to keep track of what is going on in their home country and the world. Further, Facebook pages of local institutions, shops, universities, or even asylum-specific organizations gather information on local events and happenings in an easy way (Kaufmann, 2018). One participant states: "I follow many Arabic sites, including informative and news sites. Facebook is my only way to catch up with the latest news and updates as I don't have a TV or radio. I follow these pages, and I find all the news from all over the world" (AbuJarour & Krasnova, 2017, p. 1802). This sentiment was shared by a study participant: "First in Facebook, then I will go on newspapers. I don't have TV in my current place. I'd rather do Facebook news. I'm on Twitter every day" (Marlowe, 2019, p. 179). Even though some countries try to help asylum seekers to access news, examples are German and Swedish public broadcasters who produce programs for migrants in a simpler language, these initiatives are not widely known (Graf, 2018). Another such initiative was mentioned by Tirosh and Schejter (2017): Amharic Israeli radio service (which is part of Reka, Israel Broadcast Authority (IBA) radio service for immigrants and listeners abroad) was established to help asylum seekers in Israel. However, it was also observed that quite a few participants distance themselves from news consumption (Dhoest, 2020; Kaufmann, 2018; Scheibe et al., 2019). As observed by Kaufmann (2018), some asylum seekers simply do not search for news and information. They limit their news to updates shared by family and friends, resulting in incidental news consumption. Scheibe et al. (2019) made similar observations. An interviewee explained "because there is always bad news" (p. 268) when it comes to their home country, it is preferred to abstain from news consumption.

Everyday information and tasks. As a lot of, if not all, asylum seekers' information needs are linked to their daily tasks (Alam & Imran, 2015; Coles-Kemp et al., 2018; Mansour, 2018), few different kinds of categories were observed. This ties in with the fact that we live in an information society, making it necessary to use ICT to perform everyday information activities (Díaz Andrade & Doolin, 2019). Identified information themes are related to paying utility bills (Díaz Andrade & Doolin, 2019; Witteborn, 2019), filing tax returns, preparing for the driving test, finding addresses and directions (Díaz Andrade & Doolin, 2019), banking information (Alam & Imran, 2015; Mikal & Woodfield, 2015), accessing children's school records (Mikal & Woodfield, 2015), and online shopping (Alam & Imran, 2015). Here, various kinds of apps are used to satisfy those needs, e-payment apps, location finding apps, translation apps, or apps for supporting the completion of school work (Coles-Kemp et al., 2018). Furthermore, asylum seekers perceive it of importance to obtain information of the daily lives of locals living in the new country. Some asylum seekers want to obtain information about the new country, to integrate and make life livable (Díaz Andrade & Doolin, 2019). In the case of Díaz Andrade and Doolin (2019), one study participant explained how he uses Google to find information to learn more about the new country, New Zealand, to find out how New Zealand children are taught by their parents. These information practices, finding online information, help to reduce the state of disorientation.

Mobile phones are used heavily for such tasks related to daily activities and all aspects of daily life, and also, translation of information. Not being able to translate is synonymous with not being able to exist in the new country (Coles-Kemp et al., 2018). A study by AbuJarour et al. (2019) observed that around half of the top ten used apps consist of locally used apps, including translation apps such as the Arabic-English-German dictionary arabdict. Furthermore, the prevalence of translation apps is higher among asylum seekers than immigrants. These translation apps and online dictionaries support asylum seekers in being self-dependent, for example, when they need to communicate with native speakers of the new country. A participant shared: "I use the Internet all day long. For instance, when I go to a supermarket, I write what I want to say in Arabic into the translator app, and the app translates the text to German so that the supermarket's staff know what I want to tell them" (AbuJarour & Krasnova, 2018, p. 6). The reliance on the translation of information is especially relevant in the first months after arrival (Duarte et al., 2018). "It is hard for [some of us the ones that] do not speak English because it is not Arabic copy for it" (Duarte et al., 2018, p. 8). As many asylum seekers do not speak English or, in the case of this study, German, it makes it difficult to accomplish daily tasks, for example, making a doctor's appointment (AbuJarour & Krasnova, 2017). "Here, ICTs are widely used to mitigate discomfort in communication, and help refugees achieve their goals, which enhance their sense of agency and well-being. For example, one of our respondents described how she used Google Translate to enable her to visit her physician" (AbuJarour & Krasnova, 2017, p. 1800). McCaffrey and Taha (2019) observed similar applications for translation apps, making it possible to exchange information in medical and welfare offices or even during parent-teacher conferences. Translator apps such as Google Translate or Microsoft Translator also helped to satisfy "momentary curiosities" (McCaffrey & Taha, 2019, p. 31). "Mohammad said that when his family first arrived speaking only Arabic, they wanted to know everything. 'How do you say 'sugar'?' he remembered asking his wife and kids one day in the grocery store" (McCaffrey & Taha, 2019, p. 31). Navigation apps also make up half of the top ten apps (AbuJarour et al., 2019). Named were locally used apps such as the German train navigation app *DB navigator*. Using such smartphone apps to navigate and to look up public transport schedules leads to a feeling of empowerment and self-dependence. It enables one to orient oneself in a new place and culture (Merz et al., 2018). As with navigation apps, it was observed how these apps were not necessary in the home country of asylum seekers (Duarte et al., 2018; Kaufmann, 2018). As Duarte et al. (2018) interviewed several asylum seekers, one interviewee explained "Google Earth [referring to Google Maps] I did not use it before. I did not need it before. In my land [country], I know everything. It is a small land; you do not get lost easily. Here you get lost easily" (Duarte et al., 2018, p. 12). One participant explained how he did not know about the existence of navigation apps; he only found out through a fellow asylum seeker. Another participant also disclosed:

> It was like a Sunday and I needed medicine for my father. And I didn't know, that pharmacies are closed and then I went to the pharmacy, that I know, and it was closed and then I called a friend and asked her, what [do] I have to do, and she said: 'Google which pharmacies are open in your district!', and then I googled it and found out. . . . Yeah, it comes with time, to know what you can do and then it's getting more important with time.
>
> (Kaufmann, 2018, p. 889)

This highlights the importance of sharing this kind of information with others, either through personal contacts or through social media (Alencar, 2018). GPS functions were also seen as helpful when exploring new neighborhoods and areas as well as for driving (McCaffrey & Taha, 2019). As a special kind of everyday information the topic of political engagement was observed (Leurs, 2017; Marlowe, 2019). As this is not practiced by the majority of asylum seekers, fewer studies were identified. Many refrain from voicing political opinions in fear of repercussions (Leurs, 2017). Marlowe (2019) notes:

> For these 'non-political' participants, however, it did appear that political lives did creep in over the 12 months of working with them. Whether this was in relation to a local election, concerns of what was happening in their country of origin or some other development, it became clear that adopting a political life was one of strategy and at times, necessity. What becomes evident in these comments is that political lives are at times incredibly intertwined with everyday lives. (p. 178)

Those who want to participate in political discussions in a wider context have a few strategies: "I use Facebook differently from Dutch society. I had two accounts. One political one. And another one" (Leurs, 2017, p. 688). Information is shared using virtual private network (VPN) services and using encrypted messaging services like WhatsApp, Telegram, and WickR (Leurs, 2017; Marlowe, 2019). Of course, social media is an important source for political discussion (Marlowe, 2019), for example Facebook and Instagram. Overall, according to Marlowe (2019), those who do participate in such information behavior "were unequivocal about its role in helping them and other refugees to integrate by giving them a sense of ongoing purpose" (p. 178).

Socialization

It was possible to distinguish research on how refugees maintain social ties to family and friends in their former home country and how new relationships in the new country are established. Generally speaking, different ICTs are applied as a means for communication and socialization. For example, Tirosh and Schejter (2017) observed that the most important aspect of using a computer was communication related. Here, Yahoo! Messenger was mentioned when using a computer as was observed in a study on the role of media and telecommunications in the life of asylum seekers in Israel. Similar observations were made regarding mobile phones (Ahmad, 2020); they are mostly used for communication and to maintain strong ties and to telephone family members. The Internet in general as well as social media in particular, help refugees to establish and maintain social relations (Dhoest, 2020; Köster et al., 2018). Favored social media applications for communication and socialization are WhatsApp and additionally Facebook (Mokhtar & Rashid, 2018; Scheibe et al., 2019). This need for connection is important in daily lives and WhatsApp as well as Facebook help to overcome the distance (Mokhtar & Rashid, 2018).

Keeping old contacts. When it comes to keeping in contact and socializing with relatives and friends still living in the home country or family members who had to take refuge in another country, a few studies were identified. According to AbuJarour et al. (2018) as well as Merz et al. (2018), this kind of communication is achieved mainly through ICT, which includes smartphone applications with Internet audio and video calling capabilities as well as social networking sites and various messaging applications. It was stated that the most important and first bought ICT is the telephone (mobile or smartphone) to call home (AbuJarour et al., 2018; Tudsri & Hebbani, 2015). Mobile phones are an important part in maintaining close relationships with family in the home country or who had to flee to other countries (Walker et al., 2015). This can be attributed to the fact that ICT provides cost-efficient communication. In this sample, 97.8% of the participants use the smartphone to connect to the Internet. Thus, through the means of ICT-enabled communication, the feeling of emotional support among asylum seekers can be strengthened by staying in contact with family and friends (AbuJarour et al., 2018; Bacishoga et al., 2016). For instance, one participant states how his phone was stolen and he borrowed a tablet from his friend to stay in contact with his family (Graf, 2018). This seems especially relevant for young people, as the mobile phone is used for communication and contacting family and friends—a top priority for them (Kutscher & Kreß, 2018). This helps them in insecure situations which are often characterized by different migration problems, for example uncertain legal situations. Here, Facebook, WhatsApp, Viber, and Skype were also mentioned as important social media and messaging apps.

Similar aspects can be highlighted about social media in general as it is seen as being a digital resilience resource—it makes others feel closer to each other: "Because now my family has Wi-Fi, we became able to talk with them and see them, so I feel as they live with us" (Udwan et al., 2020, p. 5). It was observed how important the social media service Facebook and messenger WhatsApp are for refugees to connect with family and friends wherever they are in the world (Alencar, 2018; Kutscher & Kreß, 2018; Mokhtar & Rashid, 2018) and how special Facebook and WhatsApp groups are used to distribute updates among family members (AbuJarour & Krasnova, 2017). Furthermore, these groups even serve as the primary source of information about family—people can share pictures and videos with family back home. Marlowe (2020) even observed how social media is mainly or even exclusively used to maintain contact with overseas family and friends. "I mean in Pakistan we may not use it that much, but because you are far from your family, and wherever the family is in the world you are in contact with them through WhatsApp, through Facebook" (Mokhtar & Rashid, 2018, p. 80). Another popular mentioned application was Skype to contact family and friends from the home country (Dhoest, 2020). Others use Skype, for example, to compare the current living circumstances and situations with relatives who were relocated in another country (Witteborn, 2019). A study on asylum seekers living in Canada shows that they use text messages as well as video and voice calls the most (Sabie & Ahmed, 2019). Another social media service that was mentioned to a smaller extent is Snapchat. A study by Leurs (2017) explained how Snapchat was used to show a new baby sister to grandparents and best friends still living in the former home country. Even YouTube is used to maintain relationships (Tudsri & Hebbani, 2015). Here, Pakistani refugees upload videos on YouTube and Facebook to stay in contact with friends and family back home. Further, to maintain old networks relies on regular communication technologies such as SMS and e-mail as well (Bacishoga et al., 2016). Here, factors such as cost, connectivity and perceived usefulness come into play. When fleeing to South Africa, in this example, SMS provided the advantage of being reasonably priced compared to voice calls. Furthermore, it is possible to send one SMS to a number of different persons. Even though mobile instant messaging (MIM) would have been a priority for the participants, it was not possible to use them since most of the people they wanted to communicate with did not have phones with MIM.

New contacts. When establishing new contacts, a few studies could be identified. Through the application of ICT and social media, a strong link between digital inclusion and social inclusion could be observed (AbuJarour & Krasnova, 2017; Alam & Imran, 2015; Alencar, 2018; Köster et al., 2018). "'Social media as a basic need for everyday life' and Participant 2, also male, said that 'if we do not have connection to each other then we will definitely get depressed . . . mental issues, psychological issues. Because we grow when we are in connection and support each other'" (Marlowe, 2020, p. 8). Using ICTs helps to feel socially included and get to know a new culture faster: "Using my smartphone and Social Media adds a lot of positivism to my life here, because I got to know the German society even better and this could help my integration process" (AbuJarour & Krasnova, 2017, p. 1800). This also works the other way—through social media locals can get to know asylum seekers better. Social media is often applied to socialize with natives of the new home country to reduce isolation (Alencar & Tsagkroni, 2019). Not being able to access the Internet was seen as disadvantageous when trying to become socially included (Alam & Imran, 2015). The younger participants stated how they felt connected to the new country by using the Internet to learn more about the new culture: "So with the help of the Internet I feel that I have been part of the Australian society. I feel integrated into the system by the Internet" (Alam & Imran, 2015, p. 356). Even though not further specified how, the smartphone is used to find other peers (AbuJarour & Krasnova, 2017). Furthermore, social media helps in finding other people of the same ethnic or religious community (Marlowe, 2020). The mobile phone is also used to communicate with people with whom no strong ties exist, e.g., work colleagues or people that were met during travel or at entertainment events (Bacishoga et al., 2016). A study by Merz et al. (2018) illustrates a point often observed in other studies (Kutscher & Kreß, 2018; Marlowe, 2020): A first personal contact with the natives of a new home country, in this example Austrians, could be successfully established online, especially through social media, rather than offline. Mentioned apps are Facebook, YouTube, LinkedIn, Twitter, Instagram, WhatsApp, Viber, and Google (Alencar, 2018). One participant states: "It's easy on Facebook to expand the circle of my friends. . . . Trust comes with time. There are people with whom I exchanged on Facebook long time before these people decide or agree that we meet in real life" (Merz et al., 2018, p. 313). A similar observation was made by Dhoest (2020). Here, one participant states: "I use it for everything: for friends, for groups, to make contact" (Dhoest, 2020, p. 21). "I could feel more confident while sharing things through social media than meeting people and talking live. I think interacting through social media could give me practice doing things [related to resettlement] with more confidence" (Marlowe, 2020, pp. 7–8). "When I came to New Zealand, I had no friends at all but now I have some friends, I found them on Facebook" (Marlowe, 2020, p. 11). Different observations were made concerning dating apps (Dhoest, 2020; Merz et al., 2018; Witteborn, 2019). In a more traditional sense, many interviewees of Dhoest (2020) use chat, dating sites, and apps to connect to other suitable partners. According to Merz et al. (2018), even dating apps, like Badoo, were used to establish contacts, not even necessarily for dating. A similar observation was made by Witteborn (2019). Here, people who fled to Hong Kong were able to meet others through dating apps. The newcomers met up with domestic workers in parks of Hong Kong on Sundays, often the only day off. Social media is also used to coordinate meetings and to see others face-to-face, such as church gatherings or activities organized by community organizations and informal gatherings. It was found that not only social media, but telephone calls, in this context intra-community calls, were indicative of bonding social capital (Koh et al., 2018).

Different groups could be contacted to socialize. Other mentioned communication means were e-mails (Díaz Andrade & Doolin, 2019). This method offers a structured and asynchronous nature of communication which helps new arriving people with language difficulties to converse in a less stressful way. Other examples were observed by Duarte et al. (2018). Social bonds can be strengthened through online platforms or groups in social media. In Münster, Germany, an App named *Welcome Münster* was developed to welcome refugees in the city. Here, one participant states: "The Welcome Münster service a lot since people helps you there . . . Germans organize parties, we see it, and we go there" (Duarte et al., 2018, p. 13). A similar initiative in the Netherlands, the "Buddy" program, was started to establish contact between newcomers and Dutch natives (Kneer et al., 2019). Here, social media served as a starting point and later to keep the connections alive. Even though the participants could not always meet face-to-face, just hanging out online is seen as a way to develop friendships. Mentioned applications were WhatsApp, Instagram, and Snapchat, whereby Snapchat was especially mentioned as a way to influence emotional and affectionate support. When offering free phones to asylum seeker families, it was observed that these were mainly used for peer support, with a secondary use for linkage with the new country's society services. Through this, the free-call phones helped to enable deepening relationships first with the peer group members and later with larger social circles (Walker et al., 2015). Bacishoga et al. (2016) looked at mobile phones from another perspective. In this study, people who arrived in South Africa met new contacts easily through similar interests at work, school, sport facilities, churches, or clinics. And through the help of mobile phones these contacts could be maintained and strengthened. Through communication with networks, of course, refugees become socially connected within the new society which correlates with aspects like agency, employment, education and language, culture, health, and many more.

Problems related to ICT and social media usage with socialization. Albeit the numerous positive effects of online media and applications, several problems and issues regarding ICT and social media in context with socialization and making contacts were described. First, Almohamed and Vyas (2019) observed how participants use social media like Facebook, Viber, or WhatsApp to contact relatives and the positive impact it has on them, but also, how they have trouble reaching those who are older. One participant explains: "But not with my mother because my mother is old, and she isn't interested to use Facebook. In my country, a small number of women has[have] an account on Facebook. Many Iraqi people believe that isn't ok for women to use Facebook" (Almohamed & Vyas, 2019,

p. 41). Even though the evidence suggests how the Internet is used to access social support, with the main purpose being support of socio-emotional and informational exchange, some studies suggest that the Internet is not used to create new relationships in the new country (Marlowe, 2020; Mikal & Woodfield, 2015). This includes either host nationals or other refugees living in the United States (Mikal & Woodfield, 2015). This is attributed to the fact that asylum seekers are reluctant to engage in online exploration or form other online communities. Some speculated reasons are cultural differences or barriers to the Internet, e.g., safety concerns, technological literacy, and access to a home Internet connection (Mikal & Woodfield, 2015). The participants seemed to meet new people face-to-face rather than online. Another interesting aspect was observed by Marlowe (2020). Here, nearly all participants (N=15) stated how social media could potentially hinder integration. As one participant put it:

> They [social media] probably hinder it because if your social media and your interactions are with people outside New Zealand, you can get a false sense that everything is OK. You don't really have to make new friends in a new land. Thus, social locations (gender, education, age) and face-to-face interactions influence people's social media encounters.
> (Marlowe, 2020, p. 8)

Another participant states: "I really don't like Facebook and I think it is wasting time. I prefer to have contact with my friends by visiting them" (Almohamed & Vyas, 2019, p. 41).

Entertainment

Concerning entertainment needs in the new country, a few studies were observed. Contradicting results were found regarding the need for entertainment and usage frequency (Ahmad, 2020; Mansour, 2018). According to Mansour (2018), only a small number of asylum seekers seemed interested in seeking entertainment. The studies by Ahmad (2020) and Merz et al. (2018) about smartphone usage for acculturation by asylum seekers determines how important the smartphone is being perceived as in keeping oneself occupied. This is also prevalent concerning entertainment, for which study participants watch movies or play apps on their devices to counteract boredom. Mentioned games that are played were Sudoku, Angry Birds, and Farmville, or listening to music, and watching videos on YouTube (Tirosh & Schejter, 2017). By exchanging funny links or jokes, entertainment needs can be satisfied with messaging services such as WhatsApp: "We sometimes exchange information on politics back home but this can lead to debates. So I have learnt to avoid those and only engage in joking or reading links related to news or entertainment" (Witteborn, 2019, p. 7). However, even being able to access entertainment programs seems to be rather surprising for some: "Coming here, you have a lot of free will to access anything, especially the internet, books, films. I had access to read books, I had access to see movies, I go to gay sites, I can watch movies online, I can watch documentaries online" (Dhoest, 2020, p. 20). The Internet is also seen as important when it comes to accessing entertainment programs in a native language (Díaz Andrade & Doolin, 2019). Examples mentioned were Nepalese or Nigerian movies, Burmese comedy, sports, music television channels, or singing karaoke. A study by Scheibe et al. (2019) cites the importance of YouTube and Facebook for entertainment purposes. Similar results were observed by Mokhtar and Rashid (2018), who applied a uses and gratifications perspective on media use by refugees. They conclude that online digital media is seen as being the most popular source to satisfy the need for entertainment. Popular sources to watch and download movies, dramas, and news videos are Facebook and YouTube. Another form of entertainment is the consumption of religious content as the preferred video and music genre (Mokhtar & Rashid, 2018; Tudsri & Hebbani, 2015). Surprisingly, only one participant cited television as a means to counteract boredom. Different insights were made by Kaufmann (2018), Mansour (2018) as well as Tudsri and Hebbani (2015). They found that indeed, many study participants spend many hours every day watching various programs on television. Programs such as sports, especially soccer, were mentioned. TV, laptops, and tablets are predominantly used for entertainment at home by the family (Kaufmann, 2018). When looking at the entertainment needs for younger asylum seekers, it is one of the most sought gratifications (Ahmad, 2020). The younger participants seem to have a higher need for this than older participants. With the highest percentage, the 15 to 17 year olds like to use their smartphone for gaming compared to the other user groups. Further, YouTube is watched a lot for entertainment and leisure purposes.

Self-presentation

Arriving at the data about the desire to present oneself online, a few studies discuss this area of interest. Self-presentation is defined as the desire to post texts, pictures, or videos of oneself or areas of interest, which include hobbies, things one created, e.g., baked goods, written messages or photography, or pictures of one's own family (Shao, 2009). According to Scheibe et al. (2019), 15 out of 19 interviewees use social media as a means to present themselves. For this, the most popular site is Facebook. Also to a minor extent, Instagram and also WhatsApp (posting a status update) are used to share pictures or information of oneself. Similar popularity was attributed to Facebook by Kutscher and Kreß (2018). Profile arrangements, postings, pictures, and quotes are all used by young asylum seekers for identity representation. By posting pictures of the home town with corresponding hashtags such as "#I_love_Syria" or creating nicknames like "free Kurdistan" on public Twitter or Instagram accounts the users can also emphasize their diasporic attachments and love for their home country, city, town, or region of origin (Leurs, 2017). This focus on cultural heritage was also observed by Dhoest (2020). Graf (2018) observed the relationship between regained self-confidence, self-presentation, and being part of an ethnic community on social media. Others, however, are careful when sharing things about oneself. One participant stated how he purposely hides some aspects of himself, for example, his sexual orientation (Dhoest, 2020). A study by Witteborn (2019) took a closer look at the displayed content on social media. It was found that often, the pictures have an upbeat tone. Observed were selfies in shopping malls, in front of global brand stores, in the streets, or in front of local attractions. Hidden are the aspects of the lives asylum seekers face such as being poor, the forced inactivity, and painful journeys. This way, the self-presentation helps to show what is meaningful to the people and not being reduced to labels like "refugee" or "asylum seeker." However, Coles-Kemp et al. (2018) made a different observation. Here, it was described that the participants did not want to convey an image of having a good time in the new country, especially to family members still living in the former home country. Therefore, a certain image of oneself is created.

8.5 Information exchange

Another focus of this literature review was the aspect of information exchange—how information is exchanged between two parties, either by face to face interaction or through online media (Wilson, 1981). Information is accessed through local authorities or colleagues and friends (Alencar & Tsagkroni, 2019; Bletscher, 2020; Nekesa Akullo & Odong, 2017; Tirosh & Schejter, 2017). According to Tirosh and Schejter (2017), the first source to access information seems to be "human rather than technological" (p. 9). These are for example members of immediate groups of the study participants—either because they have an idea what information is needed or they can refer them to the people that have the desired information. One theme that emerged when analyzing the literature was how friends or family bridge information gaps and act as information brokers

(Mansour, 2018; Mikal & Woodfield, 2015; Tirosh & Schejter, 2017). A high percentage of participants described how spoken communication with families and friends was the most important source of sought information (Mansour, 2018). To do so, a variety of technologies for communication purposes, especially mobile phones, are used.

One mentioned information area is concerned with news (Bacishoga et al., 2016; Graf, 2018; Kutscher & Kreß, 2018; Marlowe, 2020; Mokhtar & Rashid, 2018). It was observed how news is, of course, consumed through the media. However, Graf (2018) made some interesting discoveries: "It was striking in my interviews that news about the new country was primarily retrieved through the filter of the subject's own community and not directly from Swedish or German news media" (p. 155). Therefore, it can be assumed that this information exchange is focused on personal relationships and communities. It can be distinguished between worldwide news (Graf, 2018; Kutscher & Kreß, 2018) and news from the former home country, or local news (Bacishoga et al., 2016; Marlowe, 2020; Mokhtar & Rashid, 2018). When talking about local news, two thirds of the respondents participating in the study conducted by Bacishoga et al. (2016) use mobile phones to keep up with general news of their home countries and to contact friends back home to get clarity on local news. This also ties in with the importance of social media for exchanging news from the home country (Kutscher & Kreß, 2018; Marlowe, 2020; Mokhtar & Rashid, 2018).

Participants shared how they like to compare the news received from family members in the home country with news found on Facebook. "I also compare the news given by my family members from the ground to the posts shared in the Facebook. I always found that the news posted in Facebook is matched with the info received from the ground" (Mokhtar & Rashid, 2018, p. 78). If further information or more details are needed, WhatsApp is also popular: "On TV, there can be only general news, if I need some specific information for myself, like if I want some personal information, I would contact someone personally on WhatsApp, that person wants some information, they will contact me on WhatsApp" (Mokhtar & Rashid, 2018, p. 77). One participant states:

> Social media decrease the distance between New Zealand and my home country. Although it is around 24 hours by plane, I feel like nowadays we are in one home. Immediately, I can see what is happening there and I get information and their news, and they get my news. Thus, we can say that the media are very important for us nowadays.
>
> (Marlowe, 2020, p. 9)

This emphasizes the importance of social media for information exchange and news, making people feel more connected with their home country. This eradication of space and time through social media or messengers such as WhatsApp,

Viber, Facebook, or Skype, and its importance for exchanging news was observed by Kutscher and Kreß (2018) as well.

When it comes to jobs and education, a few interesting insights could be made (Alencar & Tsagkroni, 2019; Koh et al., 2018; Mansour, 2018; Stiller & Trkulja, 2018). This reliance on others in the same community is also described as helpful when searching for a job (Almohamed & Vyas, 2019): "Actually, I am relying on the Iraqi community and the Arab community to get a job because it is easy to communicate with them without English and with the cultural barriers" (p. 41). According to Koh et al. (2018), the community ties and network strength is characterized by the members looking out for each other and exchanging information. For example, one woman was able to further her education through these bonds: "There is an aged care course . . . and I didn't know about it. So one of the group give me a phone[call] . . . So this course now I will be finishing next week" (Shrestha-Ranjit et al., 2020, p. 5). This also extends to using social media for work opportunities (Alencar & Tsagkroni, 2019; Mansour, 2018). Asylum seekers are aware how important the creation of social bonds in a new home country is to generate employment opportunities (Alencar & Tsagkroni, 2019; Stiller & Trkulja, 2018).

Other information areas are concerned with health (Alencar & Tsagkroni, 2019; Shrestha-Ranjit et al., 2020), law and documents (Coles-Kemp et al., 2018), and social support (Walker et al., 2015). Health was an important aspect in the context of information exchange. This was especially relevant for interpretations done by friends or family members in context with health information. "Many participants commonly used one of their family members, usually a child, as an interpreter as they did not have another option available to them" (Shrestha-Ranjit et al., 2020, p. 5). Further, social media is applied to find and share practical health information. For example, one participant translated materials in different languages to help fellow asylum seekers (Alencar & Tsagkroni, 2019). The reliance on friends to organize certain appointments or make phone calls because of language or information deficits was described by Walker et al. (2015) and Yun et al. (2016). In context with health appointments and communication with medical staff, the participants spoke of different scenarios. One participant here states: "My daughter-in-law calls [the doctor] – she manages everything, so I really don't know how she does that all . . . My daughter-in-law takes me [to the doctor]. I cannot make it there, so my daughter-in-law takes me" (Yun et al., 2016, p. 532). Similarly: "Whenever we had any appointment, at the end we were always given an appointment, the follow-up appointment. So we're never required to make our own appointment through phone" (Yun et al., 2016, p. 532). Further, the office also offered a translation service to manage appointments. However:

> I am still scared of doing it [asking for an interpreter on the telephone] and I still feel that they might not do it for me, but I speak when somebody speaks Nepali. . . . [Health Focal Point name] used to tell me that I cannot make calls for you all the time, so you have to learn to make calls by yourself. And I'll dial the phone and then you have to speak and ask for the interpreter. (Yun et al., 2016, p. 532)

Walker et al. (2015) describe a similar situation: "I don't ask my friends to come and help me at home. But sometimes if I need to make appointment, or I need to go to hospital, I call . . . to help me out with that. Because she can speak English" (p. 331). Therefore, the language barrier can be overcome by peer support groups.

If legal regulations are concerned, an interesting story was shared: "He urgently needed a copy of her 'right to remain' in Sweden document before he would be allowed to enter Lebanon on a 24-hour visa. She quickly took a picture of the document before sending him the document; all within the space of two minutes and all done with her mobile phone" (Coles-Kemp et al., 2018, p. 5). Only through this quick exchange of information he was able to enter Lebanon. Similar observations were made about official documents and letters (Alencar, 2018; Kaufmann, 2018). Even when asylum seekers used their smartphones as translation tools, they were not able to understand official letters. Here, they relied on their new local networks and friends to translate these letters. Combining language deficits and bad digital literacy skills could especially be problematic for those who are not integrated into society. This was mirrored by Alencar (2018) as well. Here, one participant said: "As I said if I want to look for information, I will ask a friend who will help me to find the right information . . . because the content is in Dutch. So, I always go to ask a help of friends" (Alencar, 2018, p. 1596). For social support, the phone is essential to discuss problems:

> I have 5 children and my husband is sick . . . so I am really facing a financial problem . . . I get depressed . . . I contact my friend and seek advice. And so if they have good advice and a good suggestion, I will follow. Otherwise I won't. So this is a big difference in my life, that I am using the phone, and getting suggestions from my friends.
> (Walker et al., 2015, p. 330)

This highlights the importance of ICT for emotional wellbeing. Relevant information can be accessed to solve different problems. This also extends to different cultural aspects. For example:

> In my culture, being an elder you have a lot of duties that you have to do in the community, especially when it comes to marriage and stuff . . . With all these young women there are a lot of issues that the mother or someone in my age range would have been dealing with . . . if it was back home . . . And so, because of the telephone . . . these are the kinds of things that I talk sometimes with the young women.
> (Walker et al., 2015, p. 330)

This reliance was also observed in relation to children of the participants. Especially mothers seem to adopt this notion. Their children are relied on for "culture and language brokering" (Mikal & Woodfield, 2015, p. 1327). Areas in which children could help their parents were related to technology. For example, as children are often technologically more adapted and use the Internet for homework or online games, they could help their parents with the user interfaces of online job applications. This can also be extended to translation of, for example, English texts or for phone calls (McCaffrey & Taha, 2019). On the other hand, others prefer to rely on technology instead of other people. "I rely on mobile technology. And on avoiding having to ask some person, any person on the spot" (Graf, 2018, p. 153). Accordingly, the participant spoke about how he tries to avoid talking to natives of the new country in situations of risk or uncertainty and prefers to replace this interaction with the usage of an electronic device.

8.6 Conclusion

After extracting and organizing the literature, certain issues became prevalent. It was stated how insufficient access to ICT and media is seen as problematic for integration. Overall, the needed information as well as the information behavior depends on the length of stay in a new home country (Tirosh & Schejter, 2017). Several information needs identified by different studies could be determined: 1) personal: information about family members in their home country and information about how to make a living, 2) institutional: information about the status and rights as asylum seekers as well as other needs for government or municipal services, 3) spatial/orientational: concerning language, information about spatial orientation and local customs as well as information related to news from around the world (Tirosh & Schejter, 2017).

Mansour (2018) identified information needs related to work, housing, psychology, education, transportation, financial resources. Concerning ICT related practices and needs, Merz et al. (2018) mention five aspects: 1) seeking information online, 2) communicating with family and friends abroad, 3) meeting locals, 4) meeting peers, 5) counteracting boredom. Additionally, Leurs (2017) formulated five themes related to digital practices of asylum seekers: 1) right to self-determination, 2) right of freedom and expression, 3) right to information, 4) right to family, 5) right to cultural identity.

Different challenges emerge when trying to access information. When arriving in a new country, the financial means are lacking to afford technological appliances (Alam & Imran, 2015), making it difficult to gather information (Nekesa Akullo & Odong, 2017). However, smartphones are still owned by many asylum seekers, proving it to be an essential asset (Coles-Kemp et al., 2018; Sabie & Ahmed, 2019). Age differences in relation to ICT and smartphone ownership could be observed. The older the asylum seekers are, the less likely they are to use the Internet (Merisalo & Jauhiainen, 2020). Furthermore, the younger ones have more skills regarding ICT usage (Alam & Imran, 2015). This is also related to the higher access rate for the school aged children, teenagers, and young adults in contrast to the older siblings, parents, and grandparents (Bletscher, 2020). Regarding gender differences, Internet usage seems not to differ here (Merisalo & Jauhiainen, 2020). However, the ownership of smartphones seems to differ, as 40% of those who do not own a device are unmarried females between 15 to 18 years (Ahmad, 2020). Conversely, McCaffrey and Taha (2019) observed that all participants owned a phone, however, women and younger children sometimes did not have a data plan.

Possessing language skills and broader social networks in the new home country leads to less reliance on media and social media to find information (Köster et al., 2018). Further, asylum seekers seem to prefer contacting friends, family, or other people rather than using technology (Alencar & Tsagkroni, 2019; Bletscher, 2020; Nekesa Akullo & Odong, 2017; Tirosh & Schejter, 2017). However, those who have access to technology are seen as opinion leaders and therefore trusted more. Also, social media is seen as helpful with integration by developing social ties (Köster et al., 2018). Media also plays an important part in giving and receiving information, especially social media. Media is used to fulfill different needs related to entertainment, for information news/updates, for the wealth of information, and because of a positive attitude to media (Mokhtar & Rashid, 2018). In relation to this, the popular media theory – The Uses and Gratifications Theory (Katz et al., 1973) – is applicable. Media is mainly used to satisfy four needs, namely entertainment, information, socialization, and, adapted for social media, self-presentation (Shao, 2009). Concerning the aspect of information, different areas could be identified. Information related to language learning; education and job; health; law, especially asylum status; news; housing; everyday information was observed. Learning a new language in a new country is the most important concern for asylum seekers as it correlates with the desire to understand a new culture (Alencar, 2018; Merz et al., 2018). Other means are social media such as videos on Facebook, but also media of the home country such as radio broadcasts or television programs. Those who did not want to learn a new language resort to the media of their home country.

Finding education and employment is also a priority (Díaz Andrade & Doolin, 2019; Mansour, 2018). Here, ICT can be used to attend educational programs online, or find volunteer and paid employment. Further, the job market can be monitored. Additionally, for self-employed asylum seekers, customers can be recruited and retained by owning a mobile phone. Male respondents are more likely to search for information online on work opportunities than female respondents. The same applies to younger asylum seekers – they are more likely to do so than middle aged ones (Merisalo & Jauhiainen, 2019).

Regarding health information, this is one of the most important topics (Nekesa Akullo & Odong, 2017; Scheibe et al., 2019). Mothers want to know how they can access health services for their children. The Internet is used to access information on health and medications. Further, Google Translate is seen as helpful for communication when going to doctor's appointments (Almohamed & Vyas, 2019). Geolocalisation is applied to find the closest clinics. Social media is used to share experiences and give feedback regarding health procedures, e.g., on Facebook. Overall, family members with medical backgrounds are trusted the most regarding health information, even on social platforms.

Another type of information which is highly valued by asylum seekers concerns the legal system of the new home country (Alencar & Tsagkroni, 2019; Duarte et al., 2018), especially the asylum status (AbuJarour & Krasnova, 2017; Mansour, 2018). The law and documents related to this are often difficult to understand, especially if they are only available in the language of the new country. Therefore, to counteract this problem, native speakers are asked for help or information looked up on social media, especially Facebook (Köster et al., 2018). ICT is also used to translate official government websites. Furthermore, the asylum status can be looked up by accessing the official UN website.

News on different topics is also of interest, especially news about the home country (Mansour, 2018; Merisalo & Jauhiainen, 2020), but also international news. Sometimes, news channels of the new country are not accessed because of language barriers. Accessing print media in the native language of asylum seekers is often not possible, so ICT is used as an alternative to access those kinds of media. TV, online or offline radio, and news websites are accessed as well. News is consumed in English, the new countries' local language, or the native language of asylum seekers. When reading news on social media, e.g., Facebook, one has to be aware of fake news. To overcome this, information and news are compared with those received from friends and family members. Some countries do offer news channels in the native language of asylum seekers, for example Germany and Sweden, however, these programs are not widely known about. Also, some deliberately distance themselves from news, since news is often seen as too stressful.

Concerning everyday information and tasks, paying bills, filing tax returns, preparing for the driving test, finding addresses and directions, banking information, accessing children records, or online shopping are mentioned. Here, various kinds of apps are used to satisfy those needs – e-payment apps, location finding apps, translation apps, apps for supporting the completion of school work (Coles-Kemp et al., 2018). Translation apps are heavily used for daily life, it makes it possible to be self-dependent and communicate with native speakers of the new country. Going to the supermarket or a doctor's appointment or even parent teacher conferences are made easier with translation apps. These apps are also the most often installed ones on mobile phones of asylum seekers, followed by navigation apps.

The other aspect of the Uses and Gratifications Theory, socialization, was also mentioned in the literature. Differences and similarities can be observed in the way people maintain old contacts and establish new ones. Mobile phones are used to communicate and stay in contact with family members (Ahmad, 2020). Social media such as WhatsApp and Facebook are applied for communication and socialization with family back home (Mokhtar & Rashid, 2018). By staying in contact with family members and friends, a feeling of emotional support can be established. Also, SMS and e-mail are used to maintain contact as these are cheaper in contrast to phone calls. When it comes to establishing new contacts, using ICTs helps to feel socially included and get to know a new culture. Social media helps to socialize with natives of the new home country. This is also emphasized by finding other people of the same ethnic or religious community. Sometimes, it is not easy to establish contact with natives of the new home country, so social media offers a good alternative as a first contact point. Mentioned services are Facebook, Instagram, Twitter, and WhatsApp. Surprisingly, sometimes even dating apps are used to establish friendships and, of course, romantic relationships.

Some problems in relation to social media and socialization were observed. Often, it is difficult to reach the older generations. Further, there exists a reluctance overall to form new social ties via the Internet because of safety concerns, bad technological literacy, or missing access to Internet connection. Here, meeting people face-to-face is preferred. Further, relying on online contacts can hinder establishing real contacts outside of the Internet since it creates an illusion of social inclusion.

Concerning the entertainment aspect, smartphones are important in keeping oneself occupied and entertained (Ahmad, 2020; Merz et al., 2018). Watching movies and playing apps are some preferred leisure activities. Further, exchanging funny links or jokes via WhatsApp helps satisfy entertainment needs. Additionally, entertainment programs in the native language of asylum seekers can be accessed via the Internet such as movies, music channels, comedy, or sports. The younger generation seems to have the greatest need for entertainment programs (Ahmad, 2020).

Finally, the self-presentation dimension was mentioned. Facebook is used as well as Instagram and WhatsApp to post pictures or status updates of oneself. This can also be achieved by posting special hashtags or creating nicknames related to cultural origins (e.g., "#I_love_Syria," "free Kurdistan") (Leurs, 2017). Being able to represent oneself on social media is correlated with self-confidence and being part of an ethnic community. It was also observed how social media is used to create a certain image of oneself that has an upbeat tone: selfies in shopping malls, in front of global brand stores and so on. However, it was described how some purposefully do not want to convey an image of having a good time in the new country because of family members still living in the former home country.

Another important aspect related to information behavior is information exchange – how people exchange information with each other, either through face-to-face interaction or online communication. Information is accessed through local authorities or colleagues and friends (Alencar & Tsagkroni, 2019; Bletscher, 2020). The first access point seems to be human rather than technology. One area of interest is news and exchanging information about the new country – people prefer to talk about the news instead of receiving it through the new country's news channels. Further, social media and mobile phones are used to exchange information. It is possible to compare news received from family members with news found on Facebook about the home country. Through exchanging information among communities, new education and employment opportunities emerge. For example, an educational opportunity was discovered by sharing information about this. It is known how important social bonds are for employment. Other relevant information concerned with health, law and documents as well as social support were mentioned when exchanging information. Friends and family are able to translate certain documents. Further, through the application of ICT, people can feel connected and reach out for social support by phoning friends and family. Another interesting aspect concerned with information exchange is the way friends and family can act as informants. Here, social ties help to bridge information gaps and even act as information brokers. Children are often more skilled when it comes to ICT. They use technology for homework or online games and therefore are able to help parents with navigating user interfaces of online job applications. Further, they pick up the language quicker and can help with translation. In contrast, others choose to rely on technology rather than asking other people. In uncertain or risky situations, using an electronic device is preferred than asking someone on the spot.

Some problems in relation to ICT and (online) media were observed. In a new home country, it seems that some asylum seekers are introduced to new applications with which they are not familiar with (Sabie & Ahmed, 2019). This is reflected in their information behavior as well, doing things online or digitally and looking for information on the Internet is not familiar territory (Mikal & Woodfield, 2015; Tirosh & Schejter, 2017). Further, fluency of a foreign language, especially English, can impact the use of information technologies. Conversely, since one is able to use ICT, this sometimes prevents one from developing English fluency. Additionally, ICT skills are seen as important on the job market (Díaz Andrade & Doolin, 2019). These skills provide confidence and freedom (Bletscher, 2020; Merz et al., 2018). ICT is also a two sided sword – on the one hand, it provides independence because one can look up information to verify it (Merz et al., 2018). On the other hand, some seem to rely on resettlement agencies to use ICTs for them and therefore limit self-sufficiency (Bletscher, 2020). Even though the Internet is used and applied by asylum seekers, some issues were observed. Asylum seekers perceive the Internet as being dangerous. The concerns are related to frauds or being exposed to unsavory content (Coles-Kemp et al., 2018; Mikal & Woodfield, 2015; Simko et al., 2018).

References

AbuJarour, S., Bergert, C., Gundlach, J., Köster, A., & Krasnova, H. (2019). "Your home screen is worth a thousand words": Investigating the prevalence of smartphone apps among refugees in Germany. In G. Rodriguez-Abitia & C. Ferran (Eds.), *AMCIS 2019: Proceedings of the 25th Americas Conference on Information Systems* (pp. 2870–2879). Association for Information Systems.

AbuJarour, S., & Krasnova, H. (2017). Understanding the role of ICTs in promoting social inclusion: The case of Syrian refugees in Germany. In I. Ramos, V. Tuunainen, & H. Krcmar (Eds.), *ECIS 2017: Proceedings of the 25th European Conference on Information Systems* (pp. 1792–1806). Association for Information Systems.

AbuJarour, S., & Krasnova, H. (2018). E-learning as a means of social inclusion: The case of Syrian refugees in Germany. In *AMCIS 2018: Proceedings of the 24th Americas Conference on Information* (pp. 2216–2225). Association for Information Systems.

AbuJarour, S., Krasnova, H., & Hoffmeier, F. (2018). ICT as an enabler: Understanding the role of online communication in the social inclusion of Syrian refugees in Germany. In P. M. Bednar, U. Frank, & K. Kautz (Eds.), *ECIS 2018: 26th European Conference on Information Systems: Beyond Digitization – Facets of Socio-Technical Change* (pp. 23–28). Association for Information Systems.

Ahmad, M. (2020). A data analysis investigation of smart phone and social media use by Syrian refugees. *Journal of Information and Knowledge Management*, *19*(1), 1–18.

Alam, K., & Imran, S. (2015). The digital divide and social inclusion among refugee migrants: A case in regional Australia. *Information Technology and People*, *28*(2), 344–365.

Alencar, A. (2018). Refugee integration and social media: a local and experiential perspective. *Information, Communication & Society*, *21*(2), 1588–1603.

Alencar, A., & Tsagkroni, V. (2019). Prospects of refugee integration in the Netherlands: Social capital, information practices and digital media. *Media and Communication*, *7*(2), 184–194.

Almohamed, A., & Vyas, D. (2019). Rebuilding social capital in refugees and asylum seekers. *ACM Transactions on Computer-Human Interaction*, *26*(6), 1–30.

Bacishoga, K. B., Hooper, V. A., & Johnston, K. A. (2016). The role of mobile phones in the development of social capital among refugees in South Africa. *Electronic Journal of Information Systems in Developing Countries*, *72*(1), 1–21.

Bletscher, C. G. (2020). Communication technology and social integration: Access and use of communication technologies among Floridian resettled refugees. *Journal of International Migration and Integration*, *21*(2), 431–451.

Bülbül, A., & Haj Ismail, S. (2019). Refugees' social media activities in Turkey: A computational analysis and demonstration method. *Turkish Journal of Electrical Engineering and Computer Sciences*, *27*(2), 752–764.

Coles-Kemp, L., Jensen, R. B., & Talhouk, R. (2018). In a new land: Mobile phones, amplified pressures and reduced capabilities. In R. Mandryk & M. Hancock (Eds.), *CHI '18: Proceedings of the 2018 Conference on Human Factors in Computing Systems* (Article 584). Association for Computing Machinery.

Cresswell, J. W., Plano-Clark, V. L., Gutmann, M. L., & Hanson, W. E. (2003). Advanced mixed methods research designs. In T. Abbas & C. Teddlie (Eds.), *Handbook of Mixed Methods in Social & Behavioral Research* (pp. 209–240). Sage Publications.

Dhoest, A. (2020). Digital (dis)connectivity in fraught contexts: The case of gay refugees in Belgium. *European Journal of Cultural Studies*, *23*(5), 784–800.

Díaz Andrade, A., & Doolin, B. (2019). Temporal enactment of resettled refugees' ICT-mediated information practices. *Information Systems Journal*, *29*(1), 145–174.

Duarte, A. M. B., Degbelo, A., & Kray, C. (2018). Exploring forced migrants (re)settlement & the role of digital services. In *Proceedings of 16th European Conference on Computer-Supported Cooperative Work – Exploratory Papers* (pp. 1–21). European Society for Socially Embedded Technologies.

Fink, A. (2019). *Conducting research literature reviews* (5th ed.). Sage Publications.

Graf, H. (2018). Media practices and forced migration: Trust online and offline. *Media and Communication*, *6*(2), 149–157.

Katz, E., Blumler, J. G., & Gurevitch, M. (1973). Uses and gratifications research. *Public Opinion Quarterly*, *37*(4), 509–523.

Kaufmann, K. (2018). Navigating a new life: Syrian refugees and their smartphones in Vienna. *Information Communication and Society*, *21*(6), 882–898.

Kneer, J., van Eldik, A. K., Jansz, J., Eischeid, S., & Usta, M. (2019). With a little help from my friends: Peer coaching for refugee adolescents and the role of social media. *Media and Communication*, *7*(2), 264–274.

Koh, L., Walker, R., Wollersheim, D., & Liamputtong, P. (2018). I think someone is walking with me: the use of mobile phone for social capital development among women in four refugee communities. *International Journal of Migration, Health and Social Care*, *14*(4), 411–424.

Köster, A., Bergert, C., & Gundlach, J. (2018). Information as a life vest: Understanding the role of social networking sites for the social inclusion of Syrian refugees. In *39th International Conference on Information Systems: Bridging the Internet of People, Data, and Things* (pp. 4887–4896). Association for Information Systems.

Kutscher, N., & Kreß, L. M. (2018). The ambivalent potentials of social media use by unaccompanied minor refugees. *Social Media and Society*, *4*(1), 1–10.

Leurs, K. (2017). Communication rights from the margins: politicising young refugees' smartphone pocket archives. *International Communication Gazette*, *79*(6–7), 674–698.

Mansour, E. (2018). Profiling information needs and behaviour of Syrian refugees displaced to Egypt: An exploratory study. *Information and Learning Science*, *119*(3–4), 161–182.

Marlowe, J. (2019). Social media and forced migration: The subversion and subjugation of political life. *Media and Communication*, *7*(2), 173–183.

Marlowe, J. (2020). Refugee resettlement, social media and the social organization of difference. *Global Networks*, *20*(2), 274–291.

McCaffrey, K. T., & Taha, M. C. (2019). Rethinking the digital divide: Smartphones as translanguaging tools among Middle Eastern refugees in New Jersey. *Annals of Anthropological Practice*, *43*(2), 26–38.

McQuail, D. (1983). *Mass communication theory*. Sage.

Merisalo, M., & Jauhiainen, J. S. (2020). Digital divides among asylum-related migrants: Comparing internet use and smartphone ownership. *Tijdschrift voor Economische en Sociale Geografie*, *111*(5), 698–704.

Merz, A. B., Seone, M., Seeber, I., & Maier, R. (2018). A hurricane lamp in a dark night: Exploring smartphone use for acculturation by refugees. In T. Köhler, E. Schoop, & N. Kahnwald (Eds.), *Gemeinschaften in neuen Medien. Forschung zu Wissensgemeinschaften in Wissenschaft, Wirtschaft, Bildung und öffentlicher Verwaltung* (pp. 308–319). TUDpress.

Mikal, J. P., & Woodfield, B. (2015). Refugees, post-migration stress, and internet use: A qualitative analysis of intercultural adjustment and internet use among Iraqi and Sudanese refugees to the United States. *Qualitative Health Research*, *25*(10), 1319–1333.

Moher, D., Liberati, A., Tetzlaff, J., Altman, D. G., & The PRISMA Group. (2009). Preferred reporting items for systematic reviews and meta-analyses: The PRISMA statement. *PLoS Medicine*, *6*(7), Article e1000097.

Mokhtar, A., & Rashid, N. M. M. (2018). A uses and gratifications perspective on media use by refugees from Myanmar and Pakistan in Malaysia. *Al-Shajarah, Special Issue: Migration and refugee studies*, 51–105.

Nekesa Akullo, W., & Odong, P. (2017). *Information needs and information seeking behaviour of women refugees in Uganda; Public libraries' role* [Paper presentation]. LIS professionals supporting women living in conflict situations: WIL SIG Satellite Meeting, Bratislava, Slovakia.

Oduntan, O., & Ruthven, I. (2017). Investigating the information gaps in refugee integration. *Proceedings of the Association for Information Science and Technology*, *54*(1), 308–317.

Sabie, D., & Ahmed, S. I. (2019). Moving into a technology land: Exploring the challenges for the refugees in Canada in accessing its computerized infrastructures. In J. Chen, J. Mankoff, & C. Gomes (Eds.), *COMPASS '19: Proceedings of the 2nd ACM SIGCAS Conference on Computing and Sustainable Societies* (pp. 218–233). Association for Computing Machinery.

Scheibe, K., Zimmer, F., & Stock, W. G. (2019). Social media usage of asylum seekers in Germany. In W. Popma & S. Francis (Eds.), *Proceedings of the 6th European Conference on Social Media* (pp. 263–272). Academic Conferences and Publishing International.

Shao, G. (2009). Understanding the appeal of user-generated media: A uses and gratification perspective. *Internet Research*, *19*(1), 7–25.

Shrestha-Ranjit, J., Payne, D., Koziol-McLain, J., Crezee, I., & Manias, E. (2020). Availability, accessibility, acceptability, and quality of interpreting services to refugee women in New Zealand. *Qualitative Health Research*, *30*(11), 1697–1709.

Simko, L., Lerner, A., Ibtasam, S., Roesner, F., & Kohno, T. (2018). Computer security and privacy for refugees in the United States. In J. Li (Ed.), *2018 IEEE Symposium on Security and Privacy* (pp. 409–423). IEEE Computer Security.

Stiller, J., & Trkulja, V. (2018). Assessing digital skills of refugee migrants during job orientation in Germany. In G. Chowdhury, J. McLeod, V. Gillet, & P. Willett (Eds.), *Lecture notes in computer science, Vol: 10766. Transforming Digital Worlds* (pp. 527–536). Springer.

Tirosh, N., & Schejter, A. (2017). "Information is like your daily bread": The role of media and telecommunications in the life of refugees in Israel. *Hagira – Israel Journal of Migration*, *7*, 1–25.

Tudsri, P., & Hebbani, A. (2015). "Now I'm part of Australia and I need to know what is happening here": Case of Hazara male former refugees in Brisbane strategically selecting media to aid acculturation. *Journal of International Migration and Integration*, *16*(4), 1273–1289.

Udwan, G., Leurs, K., & Alencar, A. (2020). Digital resilience tactics of Syrian refugees in the Netherlands: Social media for social support, health, and identity. *Social Media and Society*, *6*(2), 1–11.

Walker, R., Koh, L., Wollersheim, D., & Liamputtong, P. (2015). Social connectedness and mobile phone use among refugee women in Australia. *Health and Social Care in the Community*, *23*(3), 325–336.

Wilson, T. D. (1981). On user studies and information needs. *Journal of Documentation*, 37(1), 3–15.

Witteborn, S. (2019). The digital gift and aspirational mobility. *International Journal of Cultural Studies*, *22*(6), 1–16.

Yun, K., Paul, P., Subedi, P., Kuikel, L., Nguyen, G. T., & Barg, F. K. (2016). Help-seeking behavior and health care navigation by Bhutanese refugees. *Journal of Community Health*, *41*(3), 526–534.

9 Identifying ICT and media practices in a new country – Age- and gender-dependent results

After fleeing from their home country, asylum seekers have to build a new life in a new country, possibly including a new information behavior and social environment. Their situation, circumstances, and living conditions are changing in the new country. Being separated from family members and friends, new social contacts need to be established. Trying to stay in touch with acquaintances and relatives is another challenge. Learning a new language for job opportunities is a requirement. When building a new life, new situations need to be adapted to: What are the needs of asylum seekers? What is their information behavior?

The following results investigate the information and communications technologies (ICTs) asylum seekers use, as well as what social media and other media they apply when settling in a new country, in this case, Germany. Describing information behavior in this study's framework, "context and situation are important concepts for information behavior research" (Case & Given, 2018, p. 48). Pettigrew, Fidel, and Bruce (2001) see "information behavior" as "how people need, seek, give, and use information in different contexts, including the workplace and everyday living" (p. 44). Accordingly, this study's approach also considers information production and dissemination behavior as well as information seeking and consumption behavior. Human information behavior is also embedded in the "information horizon" (Sonnenwald, 1999; Sonnenwald, 2005) of users and includes user's social contacts and networks. Therefore, social media and messaging services as platforms for information production and information seeking as well as reception behavior offer the basis for analysis.

Media satisfies a variety of needs by providing different kinds of information. Four central motives for applying different kinds of media – in this case social media and other online media, as well as messaging services – are defined by the Uses and Gratifications Theory (U>): information, entertainment, social interaction, and self-presentation (Blumler & Katz, 1974; McQuail, 1983; Shao, 2009). A sense of gratification is derived when using different media forms: How do asylum seekers satisfy their needs for different motivations? What kinds of media, especially social media, do they apply?

Studies suggest that generations and genders use social media differently (Barker, 2009; Fietkiewicz et al., 2016; Lenhart et al., 2010). Joiner et al. (2005) describe how women use the Internet and social media to connect with other people and communicate, whereas men like to be entertained. Looking at user-generated content by gender, men are more likely to discuss public events (e.g., politics and

https://doi.org/10.1515/9783110672022-009

sports), women tend to talk about personal stories (e.g., family related) (Wang et al., 2013). The United Nations High Commissioner for Refugees (2016) describes challenges related to connecting refugee women and the elderly as they "are less likely to have access to mobile phones and the Internet" (p. 16). Age groups are confronted with the Internet and ICT at different stages of life and therefore have more or less opportunities to adopt them (Fietkiewicz et al., 2018).

These results showcase the differences of ICT and social media usage by asylum seekers among age and gender groups. All in all, the following research questions (RQs) will be explored:

RQ2a: What ICT, online and traditional media as well as social media do asylum seekers use? Can age- and gender-dependent results be observed?

RQ2b: How do asylum seekers perceive their skills to accurately use ICT?

RQ2c: What ICT, online and traditional media as well as social media do asylum seekers apply to satisfy their needs for information, entertainment, social interaction, and self-presentation? Can age- and gender-dependent results be observed?

Semi-structured face-to-face interviews with asylum seekers were performed to gather qualitative data, and, additionally, quantitative data were collected with a questionnaire. First, as early contact points, different associations, institutions, and organizations as well as city administrations in North-Rhine Westphalia were contacted. In order to sufficiently communicate with asylum seekers, participants were selected from German language classes. Interviewed were 25 adults in Dorsten and Düsseldorf aged between 21 and 55 years in November and December 2018 as well as February 2019. All interview partners were attendees in German language courses of different levels (A1 (beginner) to C1 (proficiency)), who voluntarily took part in our interviews. Further, an Arabic native speaker was present to bridge any communication difficulties. Each interview took around 15 to 30 minutes. After analyzing the collected data, a second round of semi-structured interviews and filling in questionnaires was performed with 21 children and adolescents aged between 8 and 18 years in Wesel and Moers in April 2019 during a holiday care program. All children took part voluntarily in the interviews. In total, 45 participants (one adult was excluded as their country of origin did not fit the research focus) from Syria (N=23), Afghanistan (N=8), Turkey (N=5), Iraq (N=4), Iran (N=2), Azerbaijan (N=2), and Lebanon (N=1) were interviewed (Table 9.1). The questionnaire as well as the interview guides focused on the ICT and (social) media service usage in relation to the U> and four central motives to use media, which are *information*, *entertainment*, *social interaction*, and *self-actualization*. Self-actualization was adapted to *self-presentation* for producing content (Shao, 2009).

Table 9.1: Demographic data of interview participants; N=45.

Category	Answers	Absolute frequency	Relative frequency
Children and adolescents aged ≤18 [N=21]	Boys	14	66.7%
	Girls	7	33.3%
Adults aged >18 [N=24]	Men	14	58.3%
	Women	10	41.7%
Country [N=45]	Syria	23	51.1%
	Afghanistan	8	17.8%
	Turkey	5	11.1%
	Iraq	4	8.9%
	Azerbaijan	2	4.4%
	Iran	2	4.4%
	Lebanon	1	2.2%
Native language [N=49]. Sometimes multiple assignments	Arabic	22	44.9%
	Persian	11	22.5%
	Kurdish	9	18.4%
	Turkish	5	10.2%
	Azerbaijani	2	4.1%

First, demographics were asked about, the country of origin, age, gender, and education. Used ICTs (smartphone, Internet, TV, laptop, tablet, PC) and perceived skills to accurately use these ICT and (online) media were inquired about. Participants could rate their perceived skills on a five-point Likert-scale, 1) meaning "very bad," 3) "neither bad nor good," or 5) "very good" skills. Following, a question concerned how information (knowledge) is searched for by using social media and messaging services (e.g., WhatsApp, Facebook, YouTube, Instagram, Twitter, live streaming services) as well as online media (news websites, online encyclopedias like Wikipedia), print media (books, newspapers, magazines), or broadcasts via radio or TV. It was also inquired about what kind of information the participants are interested in (news, information about Germany, information about their home country, employment, education, legal aspects, health, religion). For the entertainment dimension, the interviewees were asked which kind of social media (see above, TikTok, Snapchat) and online media (e.g., gaming apps, streaming services like, e.g., Netflix), print media (books, newspapers, magazines), and ICT (see above) are used for fun and relaxation. Social interaction means all kinds of contact with relatives, friends, or even strangers on the Internet. Therefore, the participants were asked which social media or messaging services (WhatsApp, Facebook, Twitter, Instagram, YouTube, live streaming services, TikTok, Skype, Snapchat), or ICT (SMS, landline telephone) are used for contacting other people to satisfy the need for social connection. For the self-presentation aspect, the participants were asked which social media and messaging systems (see above) they use to share pictures, videos, or posts as well as their motives to produce online content. The interviews were transcribed by the two researchers. The texts had to be translated from German to English, whereby most sentences are reproduced analogously to their meaning.

First, some demographic data is presented, which is followed by an overview of the gender- and age-dependent differences and similarities regarding applied ICT and media, as well as the perceived skills to use ICT. As for the age distribution, two groups were formed – children and teens aged 18 and below, split between boys (N=14, 66.7%) and girls (N=7, 33.3%) as well as adults aged 19 and older divided between men (N=14, 58.3%) and women (N=10, 41.7%). This distribution matches the one of the first time asylum applicants in Germany, as around 44% of all applicants are less than 18 years old (Eurostat, 2018). Most of our participants were male (N=28, 62.2%), followed by female (N=17, 37.8%). The majority of participants came from Syria (51.1%), Afghanistan (17.8%), and Turkey (11.1%). As for education, 17 (70.8%) of the adults had a high school diploma or higher education. Of the young participants, nine (42.9%) are currently in primary school, eleven (47.6%) are in middle or high school, and two (9.5%) attend a vocational college.

9.1 Applied ICT and media

The ICT and media used by the study participants are divided between boys, girls, men, and women and are displayed in Table 9.2. Starting with the distribution among the boys, the Internet is the most used ICT (N=13, 92.86%), followed by smartphones (N=12, 85.71%). Further, the smartphone is also used to watch TV: "I do not watch movies on television, most of the time I use my smartphone or laptop for it," explained one boy (IP26). Another boy (IP37) states: "I use the Internet for school and homework." Watching TV (N=12, 85.71%), reading books (N=11, 78.57%), and using laptops (N=9, 64.29%) rank high for the young male participants as well. Half of the participants use tablets (N=7, 50.00%). Lower ranking are radios (N=5, 35.71%), PCs (N=4, 28.57%), landline telephones (N=3, 21.43%), and magazines (N=3, 21.43%). Last ranking according to the boys are newspapers (N=2, 14.29%).

Table 9.2: ICTs and media used by the study participants; N=45.

ICTs and media	Boys (N=14, 100%)		Girls (N=7, 100%)		Men (N=14, 100%)		Women (N=10, 100%)	
Smartphone	12	85.71%	7	100%	14	100%	10	100%
Laptop	9	64.29%	3	42.86%	7	50.00%	2	20.00%
PC	4	28.57%	1	14.29%	2	14.29%	1	10.00%
Tablet	7	50.00%	2	28.57%	4	28.57%	1	10.00%
TV	12	85.71%	7	100%	12	85.71%	9	90.00%
Internet	13	92.86%	7	100%	13	92.86%	10	100%
Radio	5	35.71%	3	42.86%	6	42.86%	5	50.00%
Landline	3	21.43%	2	28.57%	3	21.43%	3	30.00%
Newspaper	2	14.29%	2	28.57%	9	64.29%	9	90.00%
Magazine	3	21.43%	3	42.86%	5	35.71%	4	40.00%
Books	11	78.57%	7	100%	7	50.00%	9	90.00%

Considering the ICT and media used by the young female participants, the smartphone, the Internet, the TV as well as books are used by all participants (N=7, 100%). One girl (IP32) explains: "On the Internet I search for information for my homework." Another girl (IP35) confirms: "I use the Internet for homework, but I also like to study in the library." With a larger gap, fewer than half of the girls read magazines (N=3, 42.86%), listen to the radio (N=3, 42.86%), and use laptops (N=3, 42.86%). Newspapers, landline telephones, and tablets are each used by two (28.57%) of the young female participants. The least popular item on the list are PCs, only one of the young female participants (14.29%) uses them. "I prefer to use my smartphone instead of the computer, but I am not able to use all the functions on my smartphone," claimed one girl (IP29). Another girl states (IP28): "What can I do on laptops that I can't do on the smartphone?" highlighting the importance of smartphones for boys and girls, especially in contrast to other ICTs such as tablets and PCs.

For the adult participants, all of the men use a smartphone (N=14, 100%), followed by the Internet (N=13, 92.86%), and TV (N=12, 85.71%). "Everything is on the smartphone; on YouTube, news, information, everything," explains one man (IP1). Another man (IP4) proclaims: "You can do everything on your smartphone, why use something else." "If I do not understand a letter from the government, I use my smartphone to translate it" (IP25). Further, "I watch German TV series to learn German" (IP8). Next, newspapers are read by nine of the adult male participants (64.29%). "I read newspapers to learn German," explains one participant (IP13). Another one states: "I search for accommodations and employment in the newspaper" (IP17). With the Internet, newspapers can be read online: "I do not read newspapers or magazines, I do it on the Internet" (IP18). Many of the participants also stated how they like to read the free newspaper provided by the city they live in (IP16, IP17, IP20). "As a refugee you always have to read the newspaper" (IP25). Others simply do not find the time for media, in this case, newspapers (IP9). Ranking in the middle, books are read and laptops used by seven of the participants (50.00%). "I only read books to learn German, or the Quran," explains one man (IP1). Another participant (IP9) confirms: "I read novels in German." In the lower part of the ranking are radios (N=6, 42.86%), magazines (N=5, 35.71%), and tablets (N=4, 28.57%). "I used to listen to the radio, but not anymore" (IP4) and "I only listen to the radio in my car" (IP8, IP16). Last ranking are landline telephones (N=3, 21.43%) and PCs (N=2, 14.29%). Other participants had conflicting thoughts about smartphones and media in general: "I don't have any free time to use media, I need the time to care for my children as well as learning German," as one father states (IP7). Looking at the adult female participants, all of them use the Internet and smartphone (N=10, 100%). Almost all of the women read newspapers (N=9, 90.00%) and books (N=9, 90.00%), and watch TV (N=9, 90.00%). "I watch Syrian TV with my smartphone," one woman says (IP2). "I watch Arabic and Kurdish TV via the Internet," explains another woman (IP3). A mother clarifies: "I watch children's programs with my children" (IP11). Regarding newspapers, a woman (IP2) states: "I search for accommodation and good deals." Many of the women also read the free newspaper provided by the city (IP2, IP3, IP5, IP11). When it comes to books, a mother explains: "I read children's books with my kids" (IP11). Others use books provided by the language courses to learn German (IP12). With a bigger gap, the next ranking ICT and media are radios (N=5, 50.00%) and magazines (N=4, 40.00%). Lower ranking are landline telephones (N=3, 30.00%), laptops (N=2, 20.00%), and lastly PCs (N=1, 10.00%) and tablets (N=1, 10.00%). Half of the participants listen to the radio (N= 5, 50.00%). "I listen to the radio in my car," explains one woman (IP3). However, for some participants, there seems to be not enough time to use media. One woman states, "I do not have a lot of time for media, I need to look after my children and learn German" (IP2). "When the kids are sleeping I am able to use my smartphone and have a little bit of free time," confirms another one (IP14). "I use social media before bedtime, I need to look after my children" (IP12). And, "I only use my smartphone for 15 minutes a day, there is no time for it" (IP14).

Overall, the most popular ICTs are smartphones, the Internet, and the TV. Books are also used by all girls and almost all women and boys; in contrast, only half of the men read books. More than half of the boys also like to use laptops and tablets, which also applies to men, half of them also use laptops, but not tablets. Not many girls and women use laptops or tablets. Newspapers are read by many women and men, but boys and girls tend to not read them.

Perceived ability to use ICT and media

The participants were asked to rank their perceived ability to accurately use different ICTs on a five-point Likert-scale, 1) meaning "very bad," 3) "neither bad nor good," or 5) "very good" skills. The findings are displayed as boxplots (Figure 9.1). Boys and girls differ slightly in their perception of their skills to use ICT (median of 4, boys IQR 2 and girls IQR 1.5) with the lowest value being a 3 for both. Adult male and female participants differ quite a bit in their perception as well. Men overall display greater confidence in their abilities (median 4, IQR 2) with the lowest value at 3, whereas women are not as certain (median 3, IQR 2) with the lowest value at 2.

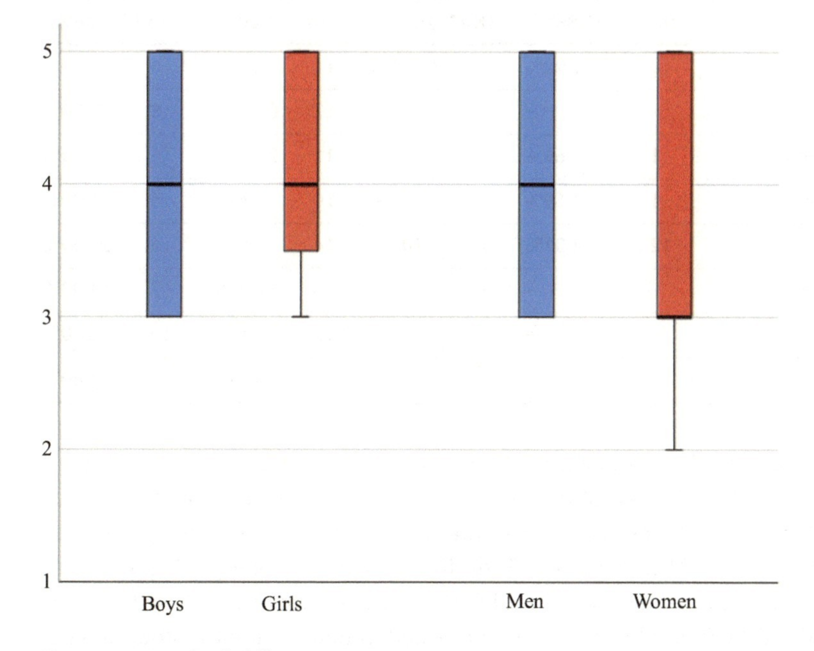

Figure 9.1: Perceived skills to use ICT; N=45.

9.2 ICT and media practices for information

The different kinds of information searched for by the participants are displayed in Table 9.3. First, the information needs of the children are described. Almost all of the boys (N=12, 92.31%) need some form of information regarding education. Here, information for school or homework are relevant (IP37). This is followed by news – ten of the boys (76.92%) are interested in them. One boy (7.69%) needs information about the home country. Information related to employment is already relevant for one of the young male participants (7.69%). Regarding the girls, all of them need information related to education such as homework (N=7, 100%). They also like to help others with homework: "I send pictures via WhatsApp about homework, if someone needs help or does not understand something" (IP29). Six of the seven girls also watch or read about news (85.71%). One girl states: "I read news in Arabic via Google" (IP28). For both juvenile participant groups, information about Germany, legal aspects, health information, and information about religion are not relevant.

Table 9.3: Information needed by the study participants; N=44.

Information category	Boys (N=13, 92.86%)		Girls (N=7, 100%)		Men (N=14, 100%)		Women (N=10, 100%)	
News	10	76.92%	6	85.71%	14	100%	8	80.00%
Germany	–	–	–	–	12	85.71%	8	80.00%
Home country	1	7.69%	–	–	11	78.57%	6	60.00%
Employment	1	7.69%	–	–	10	71.42%	5	50.00%
Education	12	92.31%	7	100%	14	100%	6	60.00%
Legal aspects	–	–	–	–	8	57.14%	7	70.00%
Health	–	–	–	–	11	78.57%	10	100%
Religion	–	–	–	–	4	28.57%	5	50.00%

Focusing on the adult participants, both, men and women, are interested in all information categories. All of the male adults need information related to education (N=14, 100%). They are all also interested in news. "I watch German and Farsi news . . . on RTL, WDR, and Todo News," explains one participant (IP1). "I watch news in German and Arabic" (IP8, IP20). Others watch ARD, WDR, ZDF, and SAT1 (IP8). While some are interested in news in their mother tongue, others try to avoid these. "No Arabic TV channels, they are all crap" (IP13) and "I don't like to listen to Arabic news" (IP25). "For a good understanding about the news you always have to consider various news portals – Russian, US-American, and different European ones – because politics are opinion-forming" (IP20). Twelve of the male participants are interested in information related to Germany (85.71%). Some want information about their home country (N=11, 78.57%). "For news about my home country I watch BBC in Farsi" (IP4). Others avoid this kind of information: "My home country is destroyed, so I am not looking for information and news about it," states one man (IP9). Health information (N=11, 78.57%) and employment (N=10, 71.42%) rank equally high. Around half of the participants (N=8, 57.14%) need information related to legal aspects. However, this aspect is also seen as difficult: "I do not search for information concerning laws or legal aspects, it is too difficult" (IP9). Another participant explains: "I received a book in German and Arabic which lists all German laws" (IP25). Not popular or as relevant are religious factors (N=4, 28.57%). "When I arrived in Germany, I asked my German supervisor about religion. However, I am not interested in it anymore" (IP9). "I am not interested in religion. It gives you a headache" (IP25). One participant also states that he is interested in information about climate change (IP13). According to the women, all of them are interested in information related to health (N=10, 100%). "I need information about health," explains one woman (IP11). Next, information about Germany as well as news are each relevant to eight (80.00%) of the ten participants. "I watch news in Kurdish, on Ronahi news sender. There are no Kurdish newspapers in Germany," explains one woman (IP2). Another one confirms: "I watch Ronahi TV via the Internet" (IP3). "I like to read and watch news in Arabic and Kurdish" (IP10). "I watch the news in Turkish and German," mentions one woman (IP22). Other TV channels mentioned are ZDF, ARD, ARTE (IP19). Legal aspects (N=7, 70.00%), information about the home country (N=6, 60.00%), and information about education (N=6, 60.00%) also rank high. Information about education does not always correlate to the participants themselves: "Information about education for my children is important to me" (IP12). Regarding information about the home country, different sentiments are expressed. "I watch Arabic news about Syria," explains one woman (IP11). However, some actively avoid information related to the home country. One woman states (IP24): "I don't want to read, see, or hear about news from my home country – it hurts me." Another woman shares this sentiment: "I do not want to know anything about my home country" (IP5). Half of the participants (N=5, 50.00%) need information related to religion and employment. "I want to know more about Christianity, also for my children. For example: What is Christmas?" (IP5). Others explain a different point of view: "Religion equals problems, so I do not inform myself about it" (IP14). Even podcasts are listened to by one woman to gather information (IP22).

All in all, adults need information on every category inquired about. Most important are news, education, and information about Germany for men, information related to health, news, and information about Germany for women. "I need information on how to live better in Germany with children," states one mother (IP5). Boys and girls are not interested in the majority of the categories. Girls and boys only need information about education and news.

Differences and similarities for the applied ICT, social media, and web services to satisfy the need for information suggested by the U> could be identified for the groups (Table 9.4). Starting with the young male participants, almost all of them use some form of media, ICT, or other service to search for or to obtain information (N=13, 92.86%). The most used services are search engines such as Google (N=13, 100%). The messaging service WhatsApp is used by ten of the boys (76.92%). Equally high ranking is YouTube (N=10, 76.92%), used to learn German or for homework. All other services are used by fewer of the participants. Instagram is used by six of the boys (46.15%) to obtain some form of knowledge. On Facebook (N=5, 38.46%), the children can ask about their friends or family. "I use Facebook to talk with my grandmother because she is in Syria" (IP40). Next, Wikipedia is used by four of the young male participants (30.77%). Last ranking are Twitter (N=2, 15.38%) and live streaming services (N=1, 7.69%). Print media such as books and magazines are not read by the boys for information or knowledge, however, one boy does read newspapers (N=1, 7.69%). ICT such as the TV and radio are each used by one participant (7.69%). Focusing on the girls, all of them use media and ICT to acquire information. Most often mentioned are WhatsApp (N=6, 85.71%) and search engines (N=6, 85.71%). YouTube is also favored by the girls to obtain information (N=5, 71.5%). For example, one girl (IP28) explains: "I learn German by watching YouTube videos." Furthermore, YouTube and Google are seen as helpful regarding homework: "For school work, I can use YouTube and Google to answer the questions," explains one girl (IP32). Facebook is used by three of the young female participants (42.86%). Used by one participant are Instagram (14.29%), Wikipedia (14.29%), and news websites (14.29%). Regarding print media, one of the girls likes to read newspapers (14.29%). Considering ICTs, one of the girls likes to watch TV (14.29%). Magazines and books are not read for information purposes by the young female participants. One girl also mentions an app for information: "I use apps for translations" (IP28).

All of the adult male participants (N=14, 100%) use media or ICT to search for information. The male participants watch videos on YouTube (N=14, 100%) for information, for example, to acquire German language skills (IP1, IP7, IP8, IP20, IP25). "YouTube can be used for everything – it has a lot of information. . . .

Table 9.4: Information dimension in relation to used services, ICT, and media; N=44.

Media, service, ICT	Boys (N=13, 92.86%)		Girls (N=7, 100%)		Men (N=14, 100%)		Women (N=10, 100%)	
WhatsApp	10	76.92%	6	85.71%	11	78.57%	9	90.00%
Facebook	5	38.46%	3	42.86%	12	85.71%	7	70.00%
Twitter	2	15.39%	–	–	5	35.71%	1	10.00%
Instagram	6	46.15%	1	14.29%	7	50.00%	5	50.00%
YouTube	10	76.92%	5	71.43%	14	100%	7	70.00%
Live streaming	1	7.69%	–	–	–	–	–	–
Search engine	13	100%	6	85.71%	14	100%	9	90.00%
News website	–	–	1	14.29%	8	57.14%	4	40.00%
Wikipedia	4	30.77%	1	14.29%	6	42.86%	2	20.00%
Newspaper	1	7.69%	1	14.29%	10	71.43%	8	80.00%
Magazine	–	–	–	–	3	21.43%	2	20.00%
Books	–	–	–	–	6	42.86%	3	30.00%
Radio	1	7.69%	–	–	5	35.71%	5	50.00%
TV	1	7.69%	1	14.29%	10	71.43%	5	50.00%

There is one person on YouTube who gives a lot of information about Germany, his name is Bamdad Esmaili [from WDR]," states one participant (IP1). Another agrees: "YouTube is better than anything else. It has all the information" (IP7). Search engines (N=14, 100%) are also used by all of the male participants to find information. "If I don't understand something, I use Google" (IP1). "I also use Google to search for news" (IP1). Further: "I use Google to learn German" (IP8). Facebook (N=12, 85.71%) and WhatsApp (N=11, 78.57%) rank second and third among the men. "I can use Facebook to watch news, mostly in English," explains one man (IP18). "I use WhatsApp groups to get information," explains another one (IP23). There are also critical voices about the validity of information found on Facebook: "There is a lot of misinformation on Facebook, you have to be careful" (IP13). Information can be exchanged in Facebook groups (IP20). News websites are read by around half of the male participants (N=8, 57.14%). "For news, I read news websites by BBC," told one participant (IP1). Following, in the middle ranks Instagram (N=7, 50.00%). Wikipedia (N=6, 42.86) is read by around half of the participants as well. Ranking last of the online services is Twitter (N=5, 35.71%). "I use Twitter to read news," explained one of the participants (IP1). Concerning print media, the majority of the male adult participants read newspapers (N=10, 71.43%) and around half of them read books (N=6, 42.86%). Magazines are less popular (N=3, 21.43%). Concerning ICT, ten (71.43%) participants watch TV and five (35.71%) listen to the radio. Considering the women, all of them use services to acquire information (N=10, 100%). One woman explains: "I am always searching for information" (IP5). WhatsApp (N=9, 90.00%) and search engines (N=9, 90.00%) are used by the majority of women. One woman (IP11) explains: "I search for cooking recipes via Google and information about health." Facebook and YouTube are used by seven (70.00%) of the women. "I use YouTube to learn German," explains one woman (IP10). This is followed by Instagram (N=5, 50.00%). Twitter is used by one (10.00%) of the adult female participants. News websites (N=4, 40.00%) and Wikipedia (N=2, 20.00%) are read by a few of the participants. One woman explains: "I read news on Kurdish news websites" (IP10). "For news, I read the *Deutsche Welle* news site," states one woman (IP22). Considering print media, the majority (N=8, 80.00%) of the women read newspapers. Books (N=3, 30.00%) and magazines (N=2, 20.00%) are read by fewer participants to acquire information. Half of the participants like to watch TV (N=5, 50.00%) and listen to the radio (N=5, 50.00%).

Live streaming services are not used by the participants, except for one boy who likes to watch live streams for information purposes. Other mentioned services were Line Live (1), Telegram (2), ebay (1) to find a flat, and Yalla Shoot (1).

9.3 ICT and media practices for entertainment

Moving on to the entertainment dimension and the used items. In order to focus on the entertainment aspect of services, ICT, and media, some of the items mentioned for the information dimension were not considered or were exchanged with other forms of media which are more related to entertainment. All of the study participants like to use media and ICT for some form of entertainment (Table 9.5). Starting with the boys, almost all of them (N=13, 92.86%) like to use apps such as playing mobile games on their smartphones to entertain themselves. This is followed by watching YouTube videos (N=12, 85.71%) for fun, such as music videos. "I use YouTube for music and games," explains one boy (IP26). Half of the boys also like to watch streaming services such as Netflix (N=7, 50.00%). However, one boy (IP31) explains: "In Turkey we watched Netflix, but now in Germany, we don't have it anymore." Instagram (N=6, 42.86%) and TikTok (N=4, 28.57%) follow next. One boy explains (IP45): "I use TikTok to make videos, but not about myself, I prefer to create them about other things." Chatting with friends or family for fun is preferred via WhatsApp (N=4, 28.57%). "I like to

Table 9.5: Entertainment dimension in relation to used services, ICT, and media; N=45.

Media, service, ICT	Boys (N=14, 100%)		Girls (N=7, 100%)		Men (N=14, 100%)		Women (N=10, 100%)	
WhatsApp	4	28.57%	3	42.86%	7	50.00%	8	80.00%
Facebook	2	14.29%	1	14.29%	10	71.43%	6	60.00%
Twitter	–	–	–	–	1	7.14%	–	–
Instagram	6	42.86%	2	28.57%	6	42.86%	2	20.00%
YouTube	12	85.71%	7	100%	13	92.86%	8	80.00%
Live streaming	2	14.29%	–	–	–	–	–	–
TikTok	4	28.57%	5	71.43%	3	21.43%	–	–
Snapchat	–	–	2	28.57%	–	–	1	10.00%
Gaming apps	13	92.86%	6	85.71%	8	57.14%	1	10.00%
Netflix	7	50.00%	–	–	4	28.57%	1	10.00%
Newspaper	–	–	–	–	3	21.43%	2	20.00%
Magazine	–	–	1	14.29%	–	–	1	10.00%
Books	1	7.14%	2	28.57%	3	21.43%	4	40.00%
Radio	1	7.14%	–	–	4	28.57%	2	20.00%
TV	4	28.57%	3	42.86%	9	64.29%	6	60.00%

post things on WhatsApp and Facebook because it is fun," explains one boy (IP36). Ranking last for the social media and web services are Facebook (N=2, 14.29%) and live streaming (N=2, 14.29%). Concerning print media, one boy (7.14%) likes to read books for fun. Watching TV (N=4, 28.57%) and listening to the radio (N=1, 7.14%) are other forms of entertainment a few of the young male participants enjoy. Not used are Twitter and Snapchat. Books and newspapers are also not read for fun. Considering the girls, all of them (N=7, 100%) like to watch videos on YouTube for fun. Also, most of them (N=6, 85.71%) use apps on their smartphones for entertainment purposes. Third ranking is watching videos on TikTok (N=5, 71.43%). Next, messaging friends on WhatsApp for fun (N=3, 42.86%), using Instagram (N=2, 28.57%), and Snapchat (N=2, 28.57%) are enjoyed by some of the young female participants. Ranking last is Facebook which is used by one (14.29%) of the girls. Concerning print media, two (28.57%) of the participants like to read books and one (14.29%) reads magazines for fun. Around half of the girls (N=3, 42.86%) like to watch TV to entertain themselves. Not used for entertainment are Twitter, live streaming, streaming services such as Netflix, the radio, and they do not read newspapers.

Moving on to the results of adult participants. Highest ranking for the men is YouTube (N=13, 92.86%), followed by Facebook (N=10, 71.43%). "I watch Afghan TV on YouTube," explains one man (IP1). "On YouTube I watch videos in Farsi, TV I watch in German" (IP4). "I like to listen to German music via YouTube," states another participant (IP23). Watching TV (N=9, 64.29%) is enjoyed by the majority of men as well as using apps on their smartphones (N=8, 57.14%) to entertain themselves. "I play PUBG on my smartphone," states one man (IP6). Another explains: "I play PUBG or Clash of Clans on my phone, but mostly I do not have time" (IP9). Interacting with friends via WhatsApp (N=7, 50.00%) is another form of entertainment enjoyed by the male adult participants. Next ranking is Instagram (N=6, 42.86%), followed by watching streaming services such as Netflix (N=4, 28.57%). Listening to the radio (N=4, 28.57%), reading books (N=3, 21.43%) and newspapers (N=3, 21.43%) as well as watching TikTok videos (N=3, 21.43%) rank in the lower half. Twitter is used by one participant (7.14%). Not used are live streaming and Snapchat, and furthermore magazines are not read by the male participants. One man (IP25) even contrasts: "In Syria, I played a lot with my smartphone, now, in Germany, I am not interested anymore. There is no time for entertainment." Focusing on the women, most of the participants like to use WhatsApp (N=8, 80.00%) and YouTube (N=8, 80.00%) for fun. Six of the adult female participants use Facebook (60.00%) and watch TV (60.00%) to entertain themselves. "I watch German, Arabic, and Kurdish TV via the Internet," explains one woman (IP3). Concerning print media, books rank the highest (N=4, 40.00%), followed by newspapers (N=2, 20.00%) and magazines (N=1, 10.00%). Last ranking for the different social media services are Instagram (N=2, 20.00%) and Snapchat (N=1, 10.00%) as well as gaming apps (N=1, 10.00%) and streaming services such as Netflix (N=1, 10.00%). Not used are Twitter, live streaming, and TikTok for the entertainment dimension. Some of the women explained that they do not use media for fun: "There is no time for entertainment, I have to look after my children" (IP3).

Overall, the boys like to use apps and watch videos for fun. The girls are similar in this regard – they additionally favor TikTok. For the men, watching videos on YouTube or TV rank high as well as using Facebook for fun. The women like to contact friends via WhatsApp or watch videos on YouTube in order to get entertained. They also like to read books. Print media is not popular for entertainment purposes overall.

9.4 ICT and media practices for socialization

Next, the socialization dimension according to the U> and the needs of the study participants are described. Here, some of the services were added and a few changed to fit the use of services in accordance with socialization (Table 9.6).

Table 9.6: Socialization dimension in relation to used services, ICT, and media; N=44.

Media, service, ICT	Boys (N=13, 92.86%)		Girls (N=7, 100%)		Men (N=14, 100%)		Women (N=10, 100%)	
WhatsApp	10	76.92%	5	71.43%	14	100%	10	100%
Facebook	2	15.38%	1	14.29%	8	57.14%	7	70.00%
Twitter	–	–	–	–	–	–	1	10.00%
Instagram	1	7.69%	–	–	5	35.71%	1	10.00%
YouTube	–	–	–	–	–	–	–	–
Live streaming	–	–	–	–	1	7.14%	–	–
TikTok	1	7.69%	–	–	1	7.14%	1	10.00%
Snapchat	1	7.69%	1	14.29%	1	7.14%	1	10.00%
Skype	–	–	–	–	1	7.14%	1	10.00%
Landline	1	7.69%	1	14.29%	3	21.43%	1	10.00%
SMS	2	15.38%	2	28.57%	6	42.86%	7	70.00%

Of the fourteen boys, thirteen (92.86%) like to use some form of media or ICT for socialization and communication. Most used to this end is WhatsApp (N=10, 76.92%). “I call my friends via WhatsApp,” states one boy (IP30). Next ranking is Facebook, two of the boys use this application (15.28%). Instagram, TikTok, and Snapchat are each used by one (7.69%) of the young male participants. One of the boys explains: “I like to post things on Facebook and Instagram to get in contact with new people” (IP34). Apart from social media, two (15.38%) of the boys write SMS, one uses the telephone (7.69%) to talk to friends and family. Twitter, YouTube, live streaming, and Skype are not used by the boys for social interaction. Moving on to the girls, all of them (N=7, 100%) like to use media for communication. Five (71.43%) use WhatsApp to this end. One girl (IP35) explains, “I like to share things on WhatsApp and Instagram because this way I can communicate with others.” Much less popular are Facebook and Snapchat, both are used by one person each (14.29%). Two of the young female participants (28.57%)

like to write SMS, one (14.29%) uses a telephone to talk with friends and family. Twitter, Instagram, YouTube, live streaming, TikTok, and Skype are not used.

Concerning the adults, all of the men (N=14, 100%) use media for socialization. Applied by all of the adult male participants is WhatsApp (N=14, 100%). One man (IP1) says: "Sometimes we are talking with video calls or audio calls via WhatsApp." "I use WhatsApp video calls to talk with others," explains one man (IP23). Around half of the male participants (N=8, 57.14%) socialize via Facebook. Following next is Instagram (N=5, 35.71%). Other used web services are live streaming platforms, TikTok, and Snapchat which are applied by one participant each. Six of the men (42.86%) also like to write SMS and three (21.43%) call others via phone or Skype (N=1, 7.14%). However, "Skype I don't use anymore, now there is Viber" (IP1). "I used to use Skype, but now there is WhatsApp" (IP21) and "Skype is so 2005" (IP25), was expressed by some participants. Not used for social interaction are Twitter and YouTube. Other social media services are mentioned: "I use Telegram for my social contacts and *imo* for video calls" (IP4). "I sometimes use imo for my social contacts" (IP8). "For my social contacts I use Facebook Messenger" (IP7). "I use *twoo* to get to know other people" (IP9). "When I play digital games with my brother I can talk to him," explains one man (IP7). For others, social media lost its appeal: "I used to interact on a lot of different social networking services, but since I am in Germany, I do not," explains one man (IP20). Others confirm "I only use WhatsApp to clarify urgent questions, not to chat or waste time" (IP25). Next, all of the women (N=10, 100%) are interested in socialization through media. WhatsApp is used by all of them (N=10, 100%). "I deleted all my social media accounts except WhatsApp to stay in contact with my family and friends," as one woman (IP24) emphasizes. Also popular is Facebook (N=7, 70.00%). Ranking last are Twitter (N=1, 10.00%), TikTok (N=1, 10.00%), Snapchat (N=1, 10.00%), and Instagram (N=1, 10.00%) regarding the media services. Skype is used by one participant only (10.00%). "I used to use Skype, but now I don't" (IP19). A majority of the women like to write SMS (N=7, 70.00%) with family and friends, one (10.00%) uses the telephone. "I write SMS if I run out of data for my mobile Internet" (IP22). Not used for socializing by women are YouTube and live streaming services.

Overall, most participants use mainly one service, in all cases WhatsApp, to socialize with friends and family. Sometimes, other services are used, mainly SMS and Facebook for the adults. The children only use WhatsApp with a few exceptions concerning other media and services.

9.5 ICT and media practices for self-presentation

Last but not least, the self-presentation dimension is shown in Table 9.7. Again, some of the services were removed to fit the use of services in accordance with self-presentation. Starting with the boys and their media practices in relation to self-presentation, six of the fourteen (42.86%) participants post information or details about themselves. Half of them (N=3, 50.00%) like to post something on Facebook and also on Instagram or TikTok. "I make videos on TikTok, but I do not show my face. My videos are about other things," explains one boy (IP40). A boy (IP34) says: "I upload pictures on Instagram and Facebook to show everybody what I am doing and to get in contact with new people." One boy states: "I post things on Facebook and WhatsApp, it is fun and I want to inform others about my situation" (IP36). Posting snaps on Snapchat (N=2, 33.33%) and status updates on WhatsApp (N=2, 33.33%) is another form of self-representation performed by the boys. One even makes videos and posts them on YouTube (16.67%). Not used are Twitter and live streaming services.

Table 9.7: Self-presentation dimension in relation to used services, ICT, and media; N=32.

Media, service, ICT	Boys (N=6, 42.86%)		Girls (N=7, 100%)		Men (N=11, 78.57%)		Women (N=8, 80%)	
WhatsApp	2	33.33%	5	71.43%	8	72.73%	4	50.00%
Facebook	3	50.00%	–	–	9	81.82%	6	75.00%
Twitter	–	–	–	–	–	–	–	–
Instagram	3	50.00%	2	28.57%	5	45.46%	2	25.00%
YouTube	1	16.67%	–	–	–	–	–	–
Live streaming	–	–	–	–	–	–	–	–
TikTok	3	50.00%	3	42.86%	2	18.18%	1	12.50%
Snapchat	2	33.33%	2	28.57%	–	–	1	12.50%

Concerning the girls, all of them (N=7, 100%) like to use media to share details about themselves. Most mentioned is WhatsApp (N=5, 71.43%). Around half of the girls (N=3, 42.86%) like to post short videos related to themselves on TikTok. They also like to post pictures on Instagram (N=2, 28.57%) or snaps on Snapchat (N=2, 28.57%). "With my friends and family I can share all my experiences via WhatsApp, I can send them pictures of my whole day," proclaims one girl (IP33). Another girl (IP42) also mentions: "I make my own videos on TikTok where I am singing. It is a lot of fun and my friends like it." Not used are Facebook, Twitter, YouTube, and live streaming by the girls.

Concerning the men, eleven (78.57%) of the fourteen participants like to post things about themselves. Most used are Facebook (N=9, 81.82%) and WhatsApp (N=8, 72.73%). Posting pictures on Instagram is enjoyed by five (45.46%) of the adult male participants. Ranking last is TikTok (N=2, 18.18%). "Posting things online and self-presentation is a lot of fun," explains one man (IP21). Self-presentation is also a form of distributing good news. "I post things about family" (IP16). A father said: "I share posts online to tell everyone that my family is fine and that we are all spending time together. It makes me happy" (IP25). Not used by them are Twitter, YouTube, live streaming, and Snapchat. Others have specific people they want to share some aspects about themselves with: "I only like to present things about myself to my friends" (IP8). "When I change my profile picture I want comments from my relatives about it" (IP20). "Self-presentation is fun when others react to it" (IP21). Others focus on special trips: "When I visit other cities I post about them" (IP17). Focusing on the women, eight of them (80.00%) post details about their lives or themselves on social media. Most used is Facebook (N=6, 75.00%) and half of the participants post status updates on WhatsApp (N=4, 50.00%). "I often change my profile picture on WhatsApp," explains one woman (IP2). Instagram (N=2, 25.00%), TikTok (N=1, 12.50%), and Snapchat (N=1, 12.50%) are used to a smaller extend by women. A woman (IP15) expresses: "On WhatsApp, Facebook, and Instagram I upload pictures, so people who know me see all the news about my life and it is fun." Some of the participants explain how they like to make posts about special occasions (IP14). A mother explains: "I post things about my family on Facebook to share special events, for example, the birthday of my daughter" (IP10). "For special occasions, for example mother's day, I post poems, or pictures of my children" (IP12). "I post pictures of my family" (IP19). Not used are Twitter, YouTube, and live streaming services.

Overall, fewer of the participants like to use media for the self-presentation dimension in contrast to the other aspects of the U>. Especially more male asylum seekers seem to avoid self-presentation. Additionally, only some of the media are used. In focus are WhatsApp, Facebook, and Instagram.

9.6 Conclusion

In conclusion, smartphones and the Internet are the most used ICTs by boys, girls, men, and women. Often utilized are the TV and books as well (however, books to a lesser extent for men). Many of the men and women also read newspapers. The least popular ICTs are PCs, tablets, and landline telephone for the adults. The PC, tablets, newspapers, and landline telephones are the least popular ICTs for the girls, and newspapers, magazines, and landline telephones for the boys. Overall, the four groups seem confident in their ability to use ICT adequately (median of 4), whereas women rate themselves lower (median of 3).

Regarding needed information, all of the women are interested in health information, followed by news and information about Germany. For the men, information about education and news has priority, followed by information about Germany. Regarding the girls, all of them need information for education such as homework. Almost all of them are also interested in news. For the boys, education and news have priority as well. However, all other categories are not relevant to the children, except for employment and information about the home country for one boy. Most used applications for boys and girls are search engines, WhatsApp, and YouTube to acquire information and knowledge, mostly related to homework or learning German. Concerning the men, they use YouTube, search engines, and Facebook the most – the women use WhatsApp, search engines, and newspapers. YouTube is especially popular to learn German and aspects about Germany. Facebook groups and WhatsApp are applied to exchange information on different topics and news. Some participants also express how they prefer to ask friends and acquaintances, especially Germans, for correct information, for example related to German traditions, the law, or politics. This can be teachers, social workers, or workers at job centers (IP5, IP10, IP13, IP16, IP17). One woman explained how her friend who lived in Germany for a few years helps her translate when going to the doctor (IP19). Another woman expressed how she trusts her husband regarding different information – about education, legal aspects, or health. Because, according to her, he is more confident in the use of the German language (IP14).

Concerning entertainment, all of the participants use media to satisfy this need. For the boys, most popular are apps on their smartphones, e.g., games, followed by YouTube, and, for half of the boys, Netflix. YouTube, gaming apps, and TikTok are used the most by girls. The children even like to make their own videos on TikTok. For the men, YouTube, Facebook, and the TV are most important for entertainment. The women like to use WhatsApp, YouTube, the TV, and Facebook for entertainment purposes. Print media are not relevant for this dimension. However, the men and women, especially mothers and a few fathers, shared how they do not use media for entertainment, as there is not enough time besides raising children, learning German, and adapting to a new country.

The socialization dimension is seen as equally important by the participants. Some even state that they deleted all social media except for WhatsApp to stay in contact with friends and family. For the boys and girls, WhatsApp is the only relevant media or messenger. Overall, they do not use a wide variety of services, messengers, or ICTs. Considering the men, all of them use WhatsApp, and most of them Facebook and SMS. The women show a similar distribution.

Moving on to the self-presentation motivation, this need is expressed by fewer of the participants. Less than half of the boys, and eleven of the fourteen men have this need. Almost all of the women and all girls like to share things related to themselves however. The boys like to post things on Facebook, Instagram, and TikTok. The girls mainly share on WhatsApp and TikTok. For the men, Facebook and WhatsApp are preferred. The women mainly use Facebook and WhatsApp as well. Fathers and mothers expressed how they like to share posts such as pictures about their family, for example about birthdays or mother's day.

Some restrictions regarding media usage were mentioned by the participants. As some services are inhibited or not as widely used in the former home country, the asylum seekers did not adopt them. For example: "Some services are restricted by the government in Iran, therefore, new ones need to be developed," as one man (IP4) expresses. Another man discloses: "In Turkey, live streaming is not allowed, therefore I do not use it . . . . We are not allowed to use Wikipedia in Turkey . . . . In Turkey, the younger generation does not use Facebook, only 'older people' are using it. It is different from Germany. Here, you need to use Facebook when you are young" (IP23). Furthermore, the most important priority for the participants is learning the new language which translates to their media use, as they watch videos on YouTube or TV to learn German. To learn a new language means being able to work and having more freedom (IP8).

References

Barker, V. (2009). Older adolescents' motivations for social network site use: The influence of gender, group identity, and collective self-esteem. *CyberPsychology & Behavior*, *12*(2), 209–213.

Blumler, J. G., & Katz, E., Eds. (1974). *The uses of mass communications: Current perspectives on gratifications research*. Sage.

Case, D. O., & Given, L. M. (2018). *Looking for information. A survey of research on information seeking, needs, and behavior* (4th ed.). Emerald Publishing Limited.

Eurostat. (2018). *Asylum statistics*. https://ec.europa.eu/eurostat/statistics-explained/index.php/Asylum_statistics#First-time_applicants:_581_thousand_in_2018

Fietkiewicz, K. J., Lins, E., Baran, K. S., & Stock, W. G. (2016). Inter-generational comparison of social media use: Investigating the online behavior of different generational cohorts. In *2016 49th Hawaii International Conference on System Sciences (HICSS)* (pp. 3829–3838). IEEE Computer Society.

Fietkiewicz, K. J., Lins, E., & Budree, A. (2018). Investigating the generation- and gender-dependent differences in social media use: A cross-cultural study in Germany, Poland and South Africa. In G. Meiselwitz (Ed.), *Lecture notes in computer science: Vol. 10914. Social Computing and Social Media. Technologies and Analytics* (pp. 183–200). Springer.

Joiner, R., Gavin, J., Duffield, J., Brosnan, M., Crook, C., Durndell, A., Maras, P., Miller, J., Scott, A. J., & Lovatt, P. (2005). Gender, internet identification, and internet anxiety: Correlates of internet use. *CyberPsychology & Behavior, 8*(4), 371–378.

Lenhart, A., Purcell, K., Smith, A., & Zickuhr, K. (2010). *Social media & mobile internet use among teens and young adults*. Pew Research Center.

McQuail, D. (1983). *Mass communication theory*. Sage.

Pettigrew, K. E., Fidel, R., & Bruce, F. (2001). Conceptual frameworks in information behavior. *Annual Review of Information Science and Technology, 35*, 43–78.

Shao, G. (2009). Understanding the appeal of user-generated media: A uses and gratification perspective. *Internet Research, 19*(1), 7–25.

Sonnenwald, D. H. (1999). Evolving perspectives of human behavior: contexts, situations, social networks and information horizons. In T. Wilson & D. Allen (Eds.), *Exploring the contexts of information behaviour* (pp. 176–190). Taylor Graham.

Sonnenwald, D. H. (2005). Information horizons. In K. E. Fisher, S. Erdelez & L. (E. F.) McKechnie (Eds.), *Theories of Information Behavior* (pp. 191–197). Information Today.

United Nations High Commissioner for Refugees. (2016). *Connecting refugees: How Internet and mobile connectivity can improve refugee well-being and transform humanitarian action*. http://www.unhcr.org/5770d43c4

Wang, Y.-C., Burke, M., & Kraut, R. (2013). Gender, topic, and audience response: An analysis of user-generated content on Facebook. In W.E. Mackay (Ed.), *CHI '13: Proceedings of the SIGCHI Conference on Human Factors in Computing Systems* (pp. 31–34). Association for Computing Machinery.

10 Home country and now – Comparing adapted media and ICT practices

To be able to answer whether asylees have changed their information behavior in the target country compared to their home country, the behavioral patterns have to be analyzed. Since the asylees' circumstances and behavior in the home country and prior to their arrival in the target country, in this case Germany, cannot be recorded through observations, asylees have to be asked about their ICT, social media and online as well as offline media usage, and motives to use these in the home country with the help of a survey.

First, participants were asked if they migrated to Germany (under normal conditions), if they were forced to migrate (asylum seeker or a refugee), or if they are a German citizen. Only surveys answered by participants who were forced to migrate were included. Afterwards, with four pre-formulated lists with social media, ICTs, online media, and offline media, the participants were asked to state which devices, services, and media they have used. The following options were given:

- social media: Facebook, Instagram, LinkedIn, live streaming services (e.g., Periscope, YouNow, Twitch, YouTube Live), messaging services (e.g., WhatsApp, Facebook Messenger, Telegram, Viber), Pinterest, Skype, Snapchat, TikTok, Twitter, YouTube, dating apps (e.g., Tinder, Badoo, Lovoo);
- ICT: Computer/notebook/laptop, television, landline telephone, smartphone, tablet;
- traditional/offline media: Books, letter/postcards, newspaper/magazine, SMS;
- digital media: Digital games (e.g., PUBG, Candy Crush), e-mail, encyclopedia (e.g., Wikipedia, Encyclopedia Britannica), learning apps (e.g., Duolingo, Babbel), podcast, search engines (e.g., Google, Bing), streaming services (e.g., Netflix, Amazon Prime, Sky Go), translation services (e.g., Google Translate, DeepL), websites of news services.

Based on the given answers, the survey attendees were asked about the usage frequency of each service, device, and media in the home country as well as in Germany. The response items were 1) "never," 2) "a few times a year," 3) "about once a month," 4) "a few times a month," 5) "about once a week," 6) "a few times a week" as well as 7) "daily" and were provided with a seven-point Likert-scale. Additionally, the motives to apply each service, device, or media in the home country as well as in Germany were asked. Here, the four specified motives 1) information (knowledge), 2) entertainment, 3) socializing (general) as well as socializing in order to be able to communicate with others, and 4) personal

https://doi.org/10.1515/9783110672022-010

identity and self-presentation from the Uses and Gratifications approach served as target items. The participants were asked, again with the help of a seven-point Likert-scale, to state if they agree or disagree to use a specific service, device, or media according to the four motives. The scale ranged from 1) "totally disagree" via 4) "neutral" to 7) "totally agree." Lastly, survey participants were asked for demographic data, e.g., their country of origin, mother tongue, age, gender, and highest educational level. Furthermore, the year of arrival in Germany, the highest German language level certificate as well as the perceived importance of having contact with and actual time spent with German citizens was asked about. An additional text field was provided in order to ask the participants for specific reasons why their behavior might have changed. There was always an option to state "I don't want to answer." The survey link was shared in social media groups for refugees and asylum seekers in order to recruit participants. However, since this did not bring any success, various organizations and institutions which volunteer to help asylum seekers or teach asylum seekers were contacted via e-mail. Thereupon, the two researchers visited integration courses and language classes to distribute the survey; the researchers and teachers of the classes were able to support participants if they needed help while answering the survey. The questionnaire could be accessed via an online link and was additionally available as a sheet of paper for those who had no access to the Internet or did not want to fill in the online version of the survey. Based on this, the following research questions (RQs) will be explored:

RQ3a: What ICT, online and traditional media as well as social media did asylum seekers use in their former home country and in Germany?

RQ3b: How frequently do or did asylum seekers use ICT, online and traditional media as well as social media in their former home country and in Germany?

RQ3c: What were the asylum seekers' motives in their former home country and in Germany to apply ICT, online and traditional media as well as social media?

A number of 69 survey participants answered all questions and were included in the data evaluation. The demographic data of the survey respondents are presented in Table 10.1. Overall, 35 survey attendees are male (50.7%) and 23 are female (33.3%) whereby 11 asylum seekers (15.9%) did not indicate their gender. The mean age of the asylum seekers and refugees is 31.9 years and the median age is 30. The youngest survey respondent is 23 years old and the oldest 54 years old. Fourteen respondents did not want to declare their age. Furthermore, with 53.6% the majority of the survey participants did not want to state their country of origin. Those who stated their country of origin are mainly from Iran (14.5%), Turkey (13.0%), and Syria (11.6%). Other asylum seekers

Table 10.1: Demographics of the survey participants.

Category	Min	Max	Median	Mean
Age [N=55]	23	54	30	31.9
			Absolute	**Relative**
Gender [N=69]	Male		35	50.7%
	Female		23	33.3%
	No answer		11	15.9%
Country [N=69]	Iran		10	14.5%
	Turkey		9	13.0%
	Syria		8	11.6%
	Iraq		2	2.9%
	Afghanistan		2	2.9%
	Azerbaijan		1	1.4%
	No answer		37	53.6%
Native language [N=69]	Arabic		20	29.0%
	Farsi		17	24.6%
	Turkish		15	21.7%
	Kurdish		7	10.1%
	Azerbaijani		2	2.9%
	Dari		2	2.9%
	Other		6	8.7%

who completed the survey are from Iraq, Afghanistan, and Azerbaijan. The most commonly mentioned mother tongues are Arabic (29.0%), Farsi (24.6%), Turkish (21.7%), and Kurdish (10.1%). Furthermore, all refugees and asylum seekers came to Germany between 2014 and 2019, the majority in the year 2015 (29.6%) followed by the year 2018 (27.8%). Nearly every surveyed asylum seeker had a German language certificate on B1 level (35.4%), B2 level (38.5%), or C1 level (18.5%). For at least 76.8% of the survey participants it is important to have contact with German citizens. Even more than half of the survey participants (N=39, 56.5%) state that it is very important to them. In line with this, 67.2% (N=64) of the refugees and asylum seekers who answered our survey indicated to have private contact with German citizens at least once a week. The participants were also asked about why they think their behavior may have changed. Many mentioned learning a new language and having to take care of the family as possible reasons.

Table 10.2 serves as an overview about which ICTs (marked red), social media (marked blue), online media (marked black), and traditional/offline media (marked green) were ever used by asylum seekers and the amount of asylum seekers which used them in the homeland and in Germany. The social media services most survey respondents have ever used in their home country or in Germany are YouTube (92.8%), Facebook (82.6%), Instagram (71.0%), and messaging services (60.9%). With a total of 92.8%, nearly every asylum seeker who took part in the survey has used the video sharing platform YouTube. Regarding ICTs, smartphones play a major role for asylum seekers as 76.8% stated they have used them. Following, the computer, notebook, or laptop (68.1%), and the television (47.8%) are used by some of the respondents as well. E-mail (85.5%) and translation services (78.3%) are the most used online media, compared to the homeland (e-mail: 63.8%; translation services: 55.1%) even more asylum seekers and refugees use them in Germany (e-mail: 82.6%; translation services: 78.3%). Search engines (68.1%) are used by most of the participants as well. Taking a look at the different offline media, around half of the asylum seekers have read books (53.6%) or written SMS (50.7%).

Table 10.2: Used social media (blue), digital media (black), traditional/offline media (green), or ICT (red).

Rank	Social media, ICT, online media, traditional media (N=69)	Absolute frequency	Relative frequency	Use in homeland	Use in Germany
1.	YouTube	64	92.8%	68.1%	81.2%
2.	E-mail	59	85.5%	63.8%	82.6%
3.	Facebook	57	82.6%	52.2%	69.6%
4.	Translation services	54	78.3%	55.1%	78.3%
5.	Smartphone	53	76.8%	73.9%	75.4%
6.	Instagram	49	71.0%	49.3%	58.0%
7.	Search engines	47	68.1%	66.7%	65.2%
	Computer/notebook/laptop	47	68.1%	63.8%	62.3%
8.	Messaging services	42	60.9%	55.1%	58.0%
9.	Books	37	53.6%	50.7%	49.3%

Table 10.2 (continued)

Rank	Social media, ICT, online media, traditional media (N=69)	Absolute frequency	Relative frequency	Use in homeland	Use in Germany
10.	SMS	35	50.7%	46.4%	43.5%
11.	Television	33	47.8%	43.5%	42.0%
12.	Encyclopedia	32	46.4%	40.6%	42.0%
13.	Streaming services	31	44.9%	18.8%	44.9%
14.	Learning apps	29	42.0%	21.7%	39.1%
	Letters/postcards	29	42.0%	17.4%	40.6%
15.	News websites	25	36.2%	33.3%	36.2%
16.	Newspapers/magazines	24	34.8%	31.9%	30.4%
17.	Digital games	19	27.5%	24.6%	23.2%
18.	Twitter	18	26.1%	21.7%	21.7%
19.	Live streaming services	16	23.2%	17.4%	20.3%
	Tablet	16	23.2%	17.4%	18.8%
20.	Skype	15	21.7%	17.4%	14.5%
	Snapchat	15	21.7%	8.7%	20.3%
21.	LinkedIn	11	15.9%	8.7%	14.5%
	Landline telephone	11	15.9%	14.5%	5.8%
22.	Podcast	10	14.5%	8.7%	14.5%
23.	TikTok	9	13.0%	4.3%	11.6%
24.	Pinterest	7	10.1%	7.2%	8.7%
25.	Dating apps	5	7.2%	1.4%	7.2%

Many asylum seekers who did not use social media services or online media in their home country began using them in Germany. More asylees use social media like YouTube, Facebook, Instagram or Snapchat, TikTok, LinkedIn, and even dating apps, now, in Germany. Taking a look at online media, especially e-mail and translation services are used by more asylum seekers since living in Germany, as well as streaming services like Netflix or learning apps. For offline media, only letters are written by more asylum seekers in Germany than in the home country. Regarding other offline media and also for ICTs, the amount of asylum seekers who use them since living in Germany is nearly the same. The only media, device, or service the majority of asylees stopped using since living in Germany is the landline telephone.

In the following, the data displaying the asylum seekers' motives to apply a particular service, media, or device in the home country and in Germany are presented as diverging stacked bar charts showing how much respondents agreed, disagreed, or had a neutral opinion. For better readability, the term "home country" or "homeland" was chosen, meaning the study participants' former country of usual residence. The usage frequencies of a service, media, or device in the homeland and in Germany are presented with normal stacked bar charts. The N-values display the total number of answers given for each question and not the related N-values corresponding to the given percentages. The next paragraphs are very number-oriented and also very detailed. Readers not interested in such details may skip those paragraphs and continue reading the conclusion paragraph.

10.1 Social media services

YouTube

Most of the respondents use YouTube in Germany daily (77.19%; N=57) or a few times a week (15.79%). As shown in Figure 10.1, for the usage frequency in the asylum seekers' homeland, only 39.29% (N=56) stated they have used YouTube on a daily basis and 10.71% have used YouTube a few times per week. Furthermore, the data shows that 16.07% did not use YouTube in their home country. Accordingly, YouTube is used more frequently by asylum seekers in Germany compared to the usage in their home country.

Looking at the motives of asylum seekers to use YouTube, information and entertainment are by far the most important aspects to use the video sharing platform, for both, the homeland and Germany alike. While 63.62% of the respondents agreed that they have used YouTube for information, 84.90% agreed that they use YouTube in Germany for the purpose of obtaining information, resulting in an increase of 21.28 percent points. Similarities can be found for entertainment as well, while 62.79% stated they agree they have used YouTube for entertainment, 76.00% (13.21 percent points increase) agreed to use YouTube for entertainment in Germany. Socializing and self-presentation are not a motive to use YouTube for most of the asylum seekers. More than half of the respondents disagreed that they use YouTube for socializing or self-presentation in Germany or disagreed they have used YouTube for socializing or self-presentation in their home country.

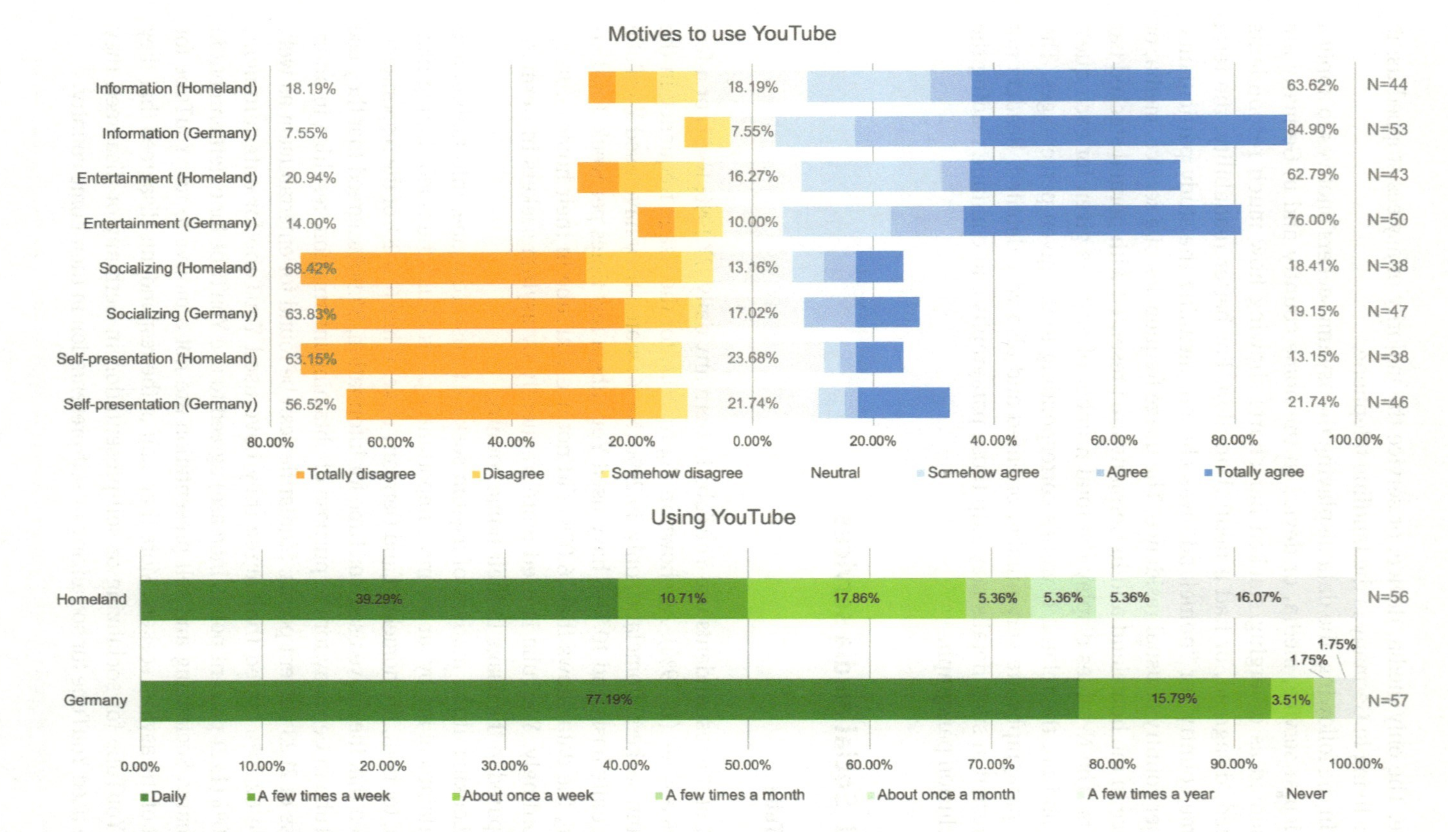

Figure 10.1: Motives to use YouTube in home country and Germany as well as usage frequency.

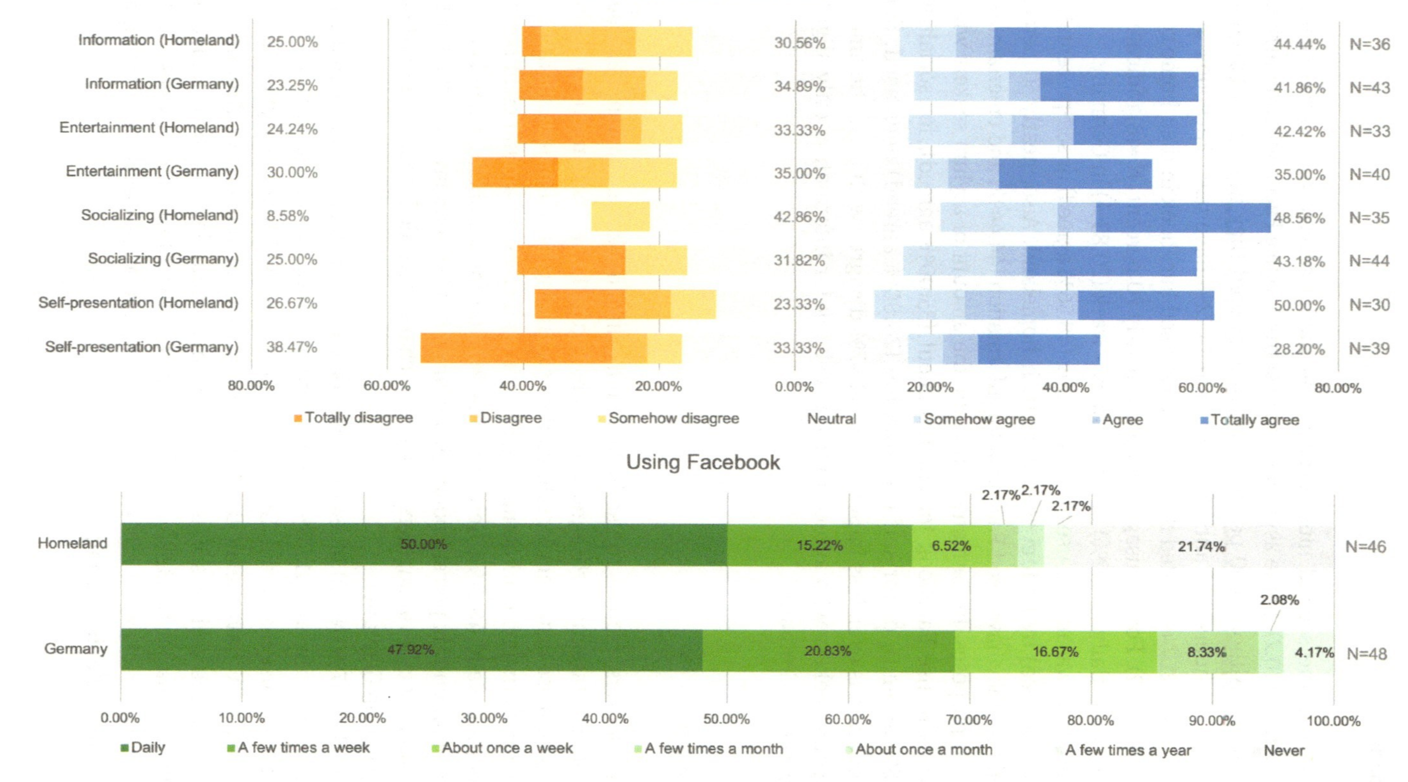

Figure 10.2: Motives to use Facebook in home country and Germany as well as usage frequency.

Facebook

Facebook is the second most used social media service by asylum seekers (82.6%; Table 10.2). Some asylum seekers started to use Facebook after arriving in Germany. Although 50 percent (N=46) answered they have used Facebook daily in their homeland and an additional 15.22% have used Facebook a few times a week, 21.74% stated they never have used Facebook in their home country (Figure 10.2). For the usage frequency, nearly 70 percent (N=48) indicated to use Facebook daily (47.92%) or a few times a week (20.83%) in Germany.

Facebook was mainly used for socializing (48.56% agreed) or self-presentation (50.00% agreed) by asylum seekers in their homeland. Also, nearly 50 percent agreed that they have used Facebook for information (44.44%) and entertainment (42.42%) purposes. In Germany, especially the amount of asylum seekers whose motive is to use Facebook for entertainment (35.00% agreed), for socializing (43.18% agreed), or for self-presentation (28.80% agreed) decreased. Self-presentation is even less important as 38.47% disagree with this. The amount of asylum seekers who agreed to use Facebook for information only slightly decreased in Germany.

Instagram

Asylum seekers' motives to use Instagram are information, socializing, and self-presentation (around 50% agreed for each; Figure 10.3). The amount of asylum seekers who use Instagram for information, socializing, or self-presentation has increased in Germany. For information, there is an increase of around eight percent points (homeland 48.27%; Germany 56.42%) and for socializing there is an increase of around six percent points (homeland 48.38%; Germany 54.05%). For the entertainment motivation, there is nearly no change recorded (decrease of 1.67 percent points in Germany).

Total 78.57% (N=42) of asylum seekers who use Instagram do so on a daily basis in Germany and 11.90% use Instagram a few times a week. Only 2.38% stated to not use Instagram. In comparison, 8.11% (N=37) of the asylum seekers indicated they have never used Instagram in their homeland. Around half (51.35%) of the asylum seekers have used Instagram daily in their homeland, nearly one quarter (24.32%) stated to have used it a few times a week, 5.41% about once a week, and 10.81% a few times a month. A clear increase of Instagram use is noticeable since living in Germany.

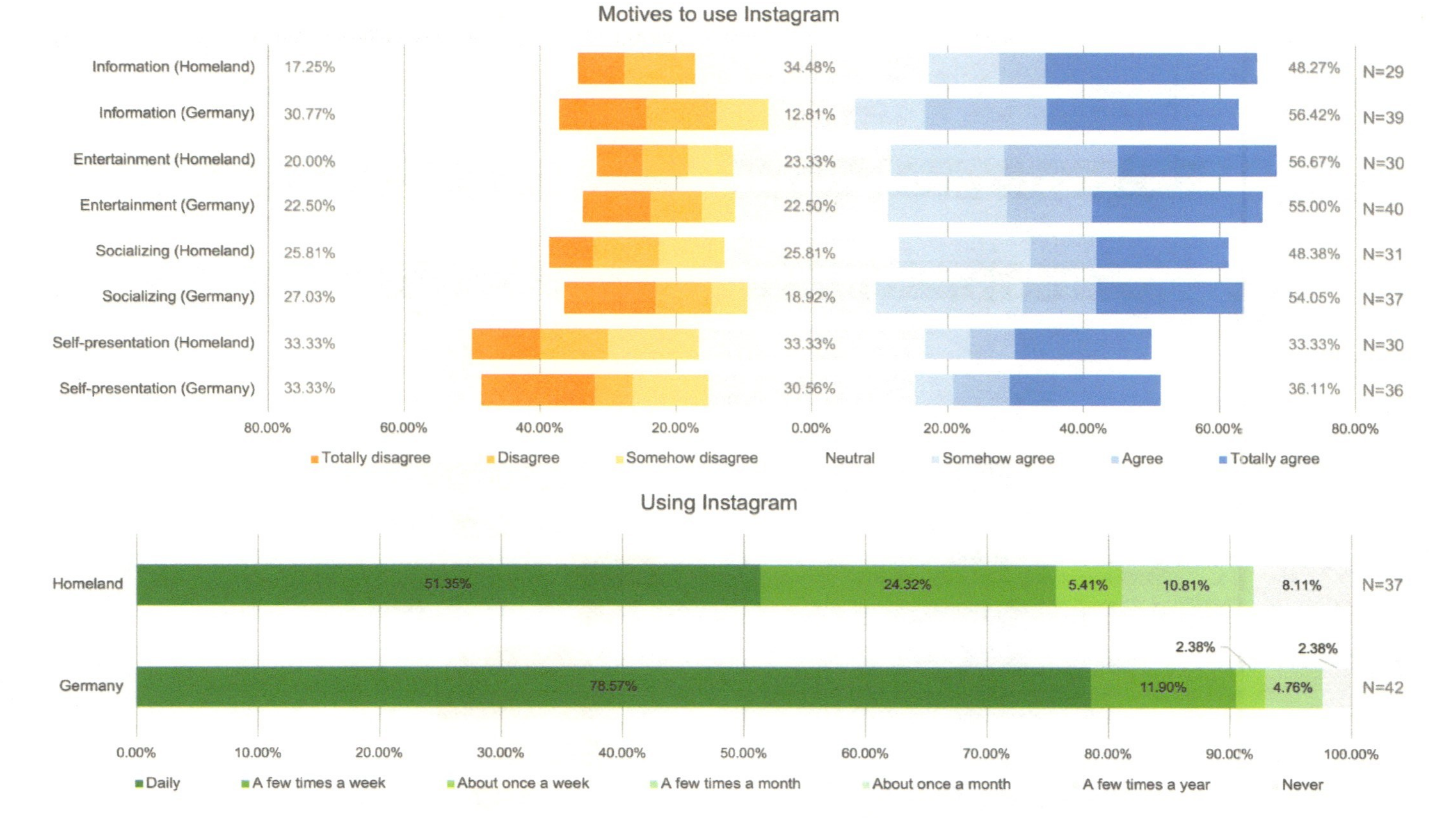

Figure 10.3: Motives to use Instagram in home country and Germany as well as usage frequency.

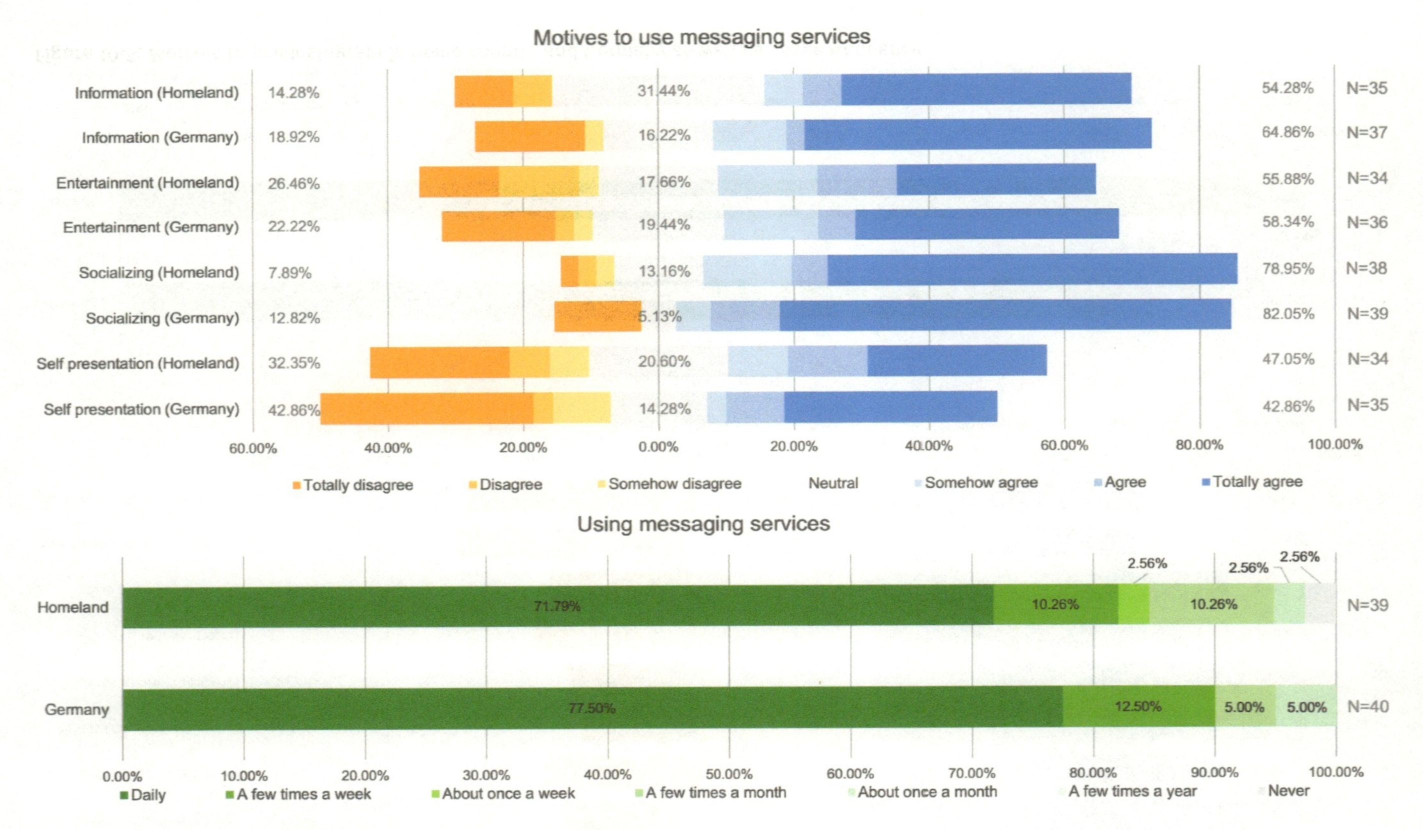

Figure 10.4: Motives to use messaging services in home country and Germany as well as usage frequency.

Messaging services

Messaging services, including e.g., WhatsApp, Facebook Messenger, Telegram, and Viber, are used very frequently by asylum seekers. Around 70 percent (71.79%; N=39) have used them daily in their homeland, displayed in Figure 10.4. In Germany, 90 percent (N=40) of those who use messaging services do so daily (77.50%) or a few times a week (12.50%). Furthermore, everyone who answered uses these services at least once a month in Germany. A few (2.56%) of the responding asylum seekers use messaging services now and did not use them in their home country.

Nearly 80 percent (78.95%) have used messaging services for socializing in their home country. In Germany, the amount of asylum seekers who are using messaging services to socialize increased (82.05% agreed). Also, more than 50 percent of asylum seekers agreed to use messaging services for information. The number of asylum seekers who reported using messaging services for information increased by around 10 percent points (homeland 54.28%; Germany 64.86%). It is noticeable that 42.86% disagreed and 42.86% agreed to use messaging services for self-presentation in Germany. An amount of 32.35% disagreed and 47.05% agreed they have used messaging services for self-presentation purposes in their homeland.

Twitter

Looking at the motives to use the micro-blogging service Twitter, it strikes that the main motive is to receive information (Figure 10.5). Total 71.42% of asylum seekers and refugees are motivated to use it for information in Germany. The amount of asylum seekers who agreed to be motivated to use Twitter for information purposes in Germany increased by around 20 percent points in comparison to the home country (from 50.00% to 70.00%). For the motives entertainment, socializing, and self-presentation the majority disagreed to be motivated by it, for Germany even more than for the homeland. In particular, 61.53% disagreed to use Twitter for self-presentation in Germany.

Regarding the usage frequency of Twitter in the asylum seekers' homeland as well as the usage frequency in Germany, 12.50% (N=16 each) indicated to never use the service. Asylees use Twitter more frequently in Germany. Total 37.50% have used Twitter on a daily basis; now, in the target country, it is 43.75%.

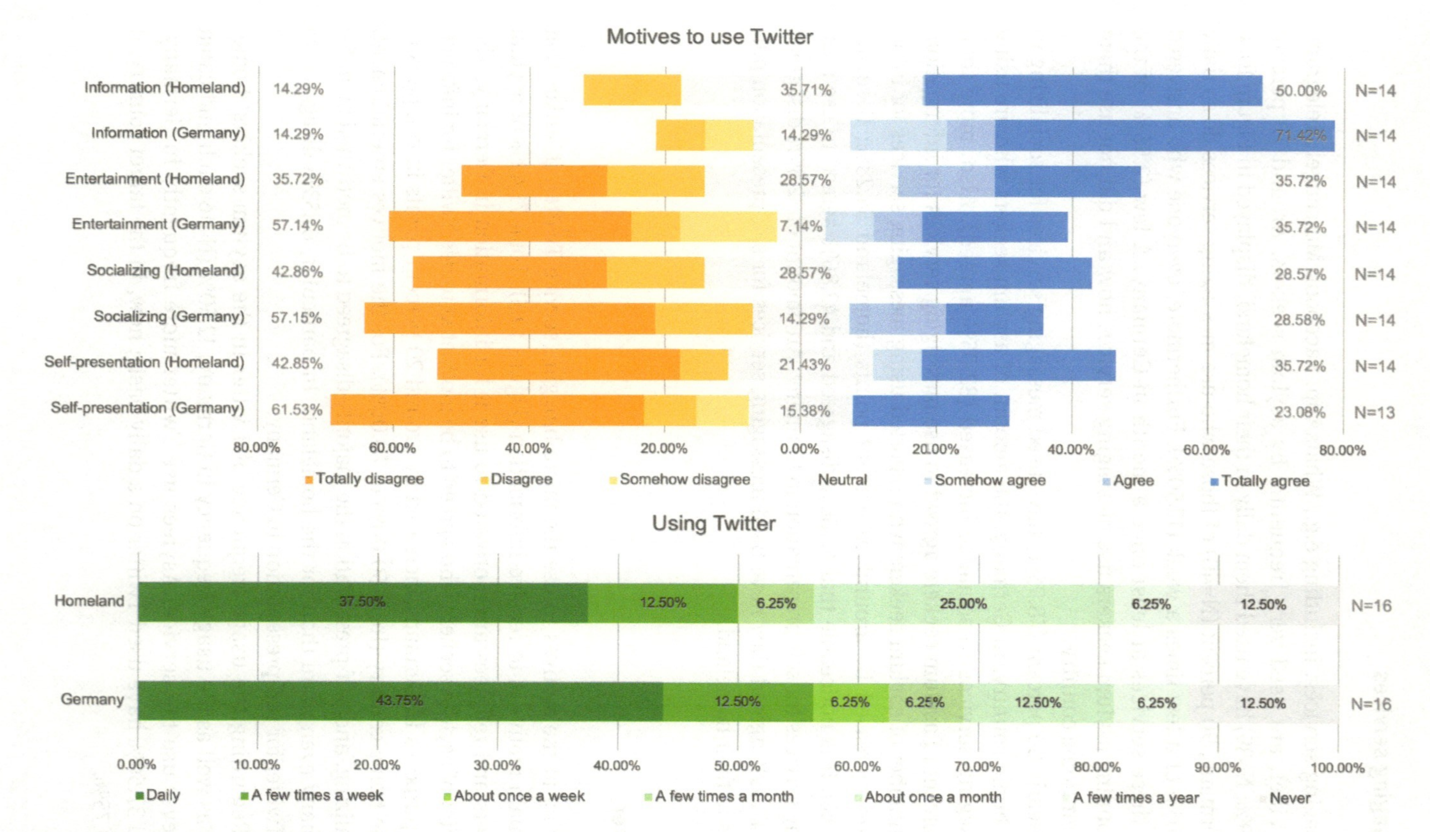

Figure 10.5: Motives to use Twitter in home country and Germany as well as usage frequency.

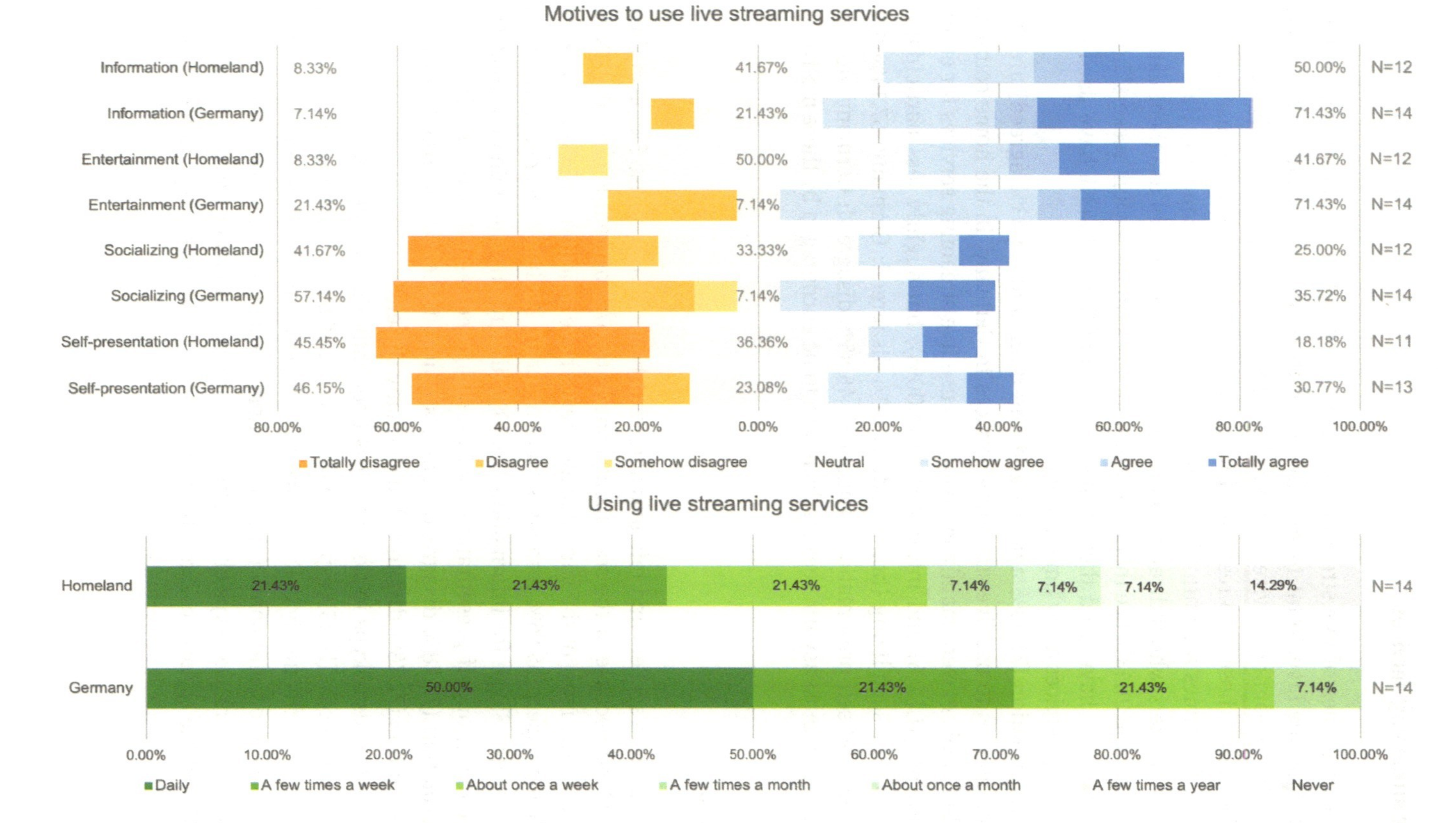

Figure 10.6: Motives to use live streaming services in home country and Germany as well as usage frequency.

Live streaming services

Some examples of live streaming services are YouNow, Taobao Live, Twitch, and YouTube Live. Total 14.29% (N=14) of the survey respondents who have used live streaming services did not do so in their home country (Figure 10.6). In Germany, those who responded to use live streaming services, use them at least a few times a month (7.14%; N=14) or even more frequently. Half of the asylum seekers who indicated to use live streaming services use them daily in Germany and 21.43% use them a few times a week. In the asylum seekers' homeland, 21.43% have used them daily and an additional 21.43% have used live streaming services a few times a week.

Information (71.43% agreed) and entertainment (71.43% agreed) are the main motives to use live streaming services in Germany. In the home country, asylum seekers were mainly motivated to use live streaming services for entertainment (41.67%) and information (50.00%), but slightly less participants agreed. For socializing (homeland: 41.67% disagreed; Germany: 57.14% disagreed) and self-presentation (homeland: 45.45% disagreed; Germany: 46.15% disagreed), most asylum seekers disagreed to be encouraged by these motives.

Skype

Most asylum seekers and refugees agreed to be motivated to use Skype in their homeland because of the socializing aspect (66.67% agreed; Figure 10.7). Now in Germany, most of the survey respondents have a neutral opinion on the socializing aspect (55.56% responses were neutral). Asylum seekers and refugees did not use Skype for information (75.00% disagreed), entertainment (66.67% disagreed), or self-presentation (72.73% disagreed) and do not use Skype for information (77.78% disagreed), entertainment (66.67% disagreed), or self-presentation (77.78% disagreed) in Germany either.

Asylees use Skype in Germany less frequently than in the homeland. While nearly 70 percent (N=15) responded they have used Skype once a month (26.67%) or even more frequently in their home country, only around 35 percent (N=14) use Skype about once a month (14.29%) or more frequently in Germany. Most of the asylum seekers use Skype only a few times a year in Germany (36.71%), and even 28.57% of the asylum seekers who indicated they have used Skype in their homeland, do not use it in Germany.

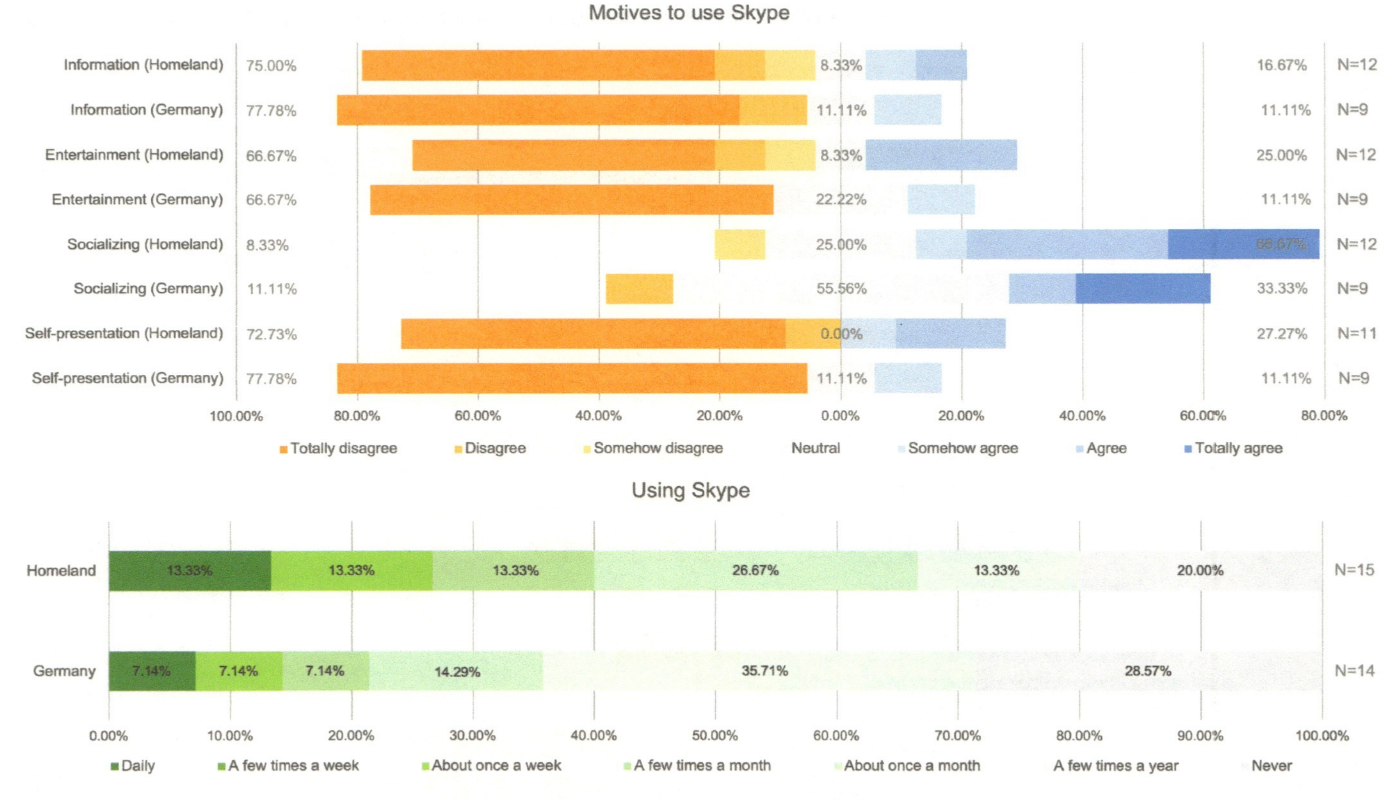

Figure 10.7: Motives to use Skype in home country and Germany as well as usage frequency.

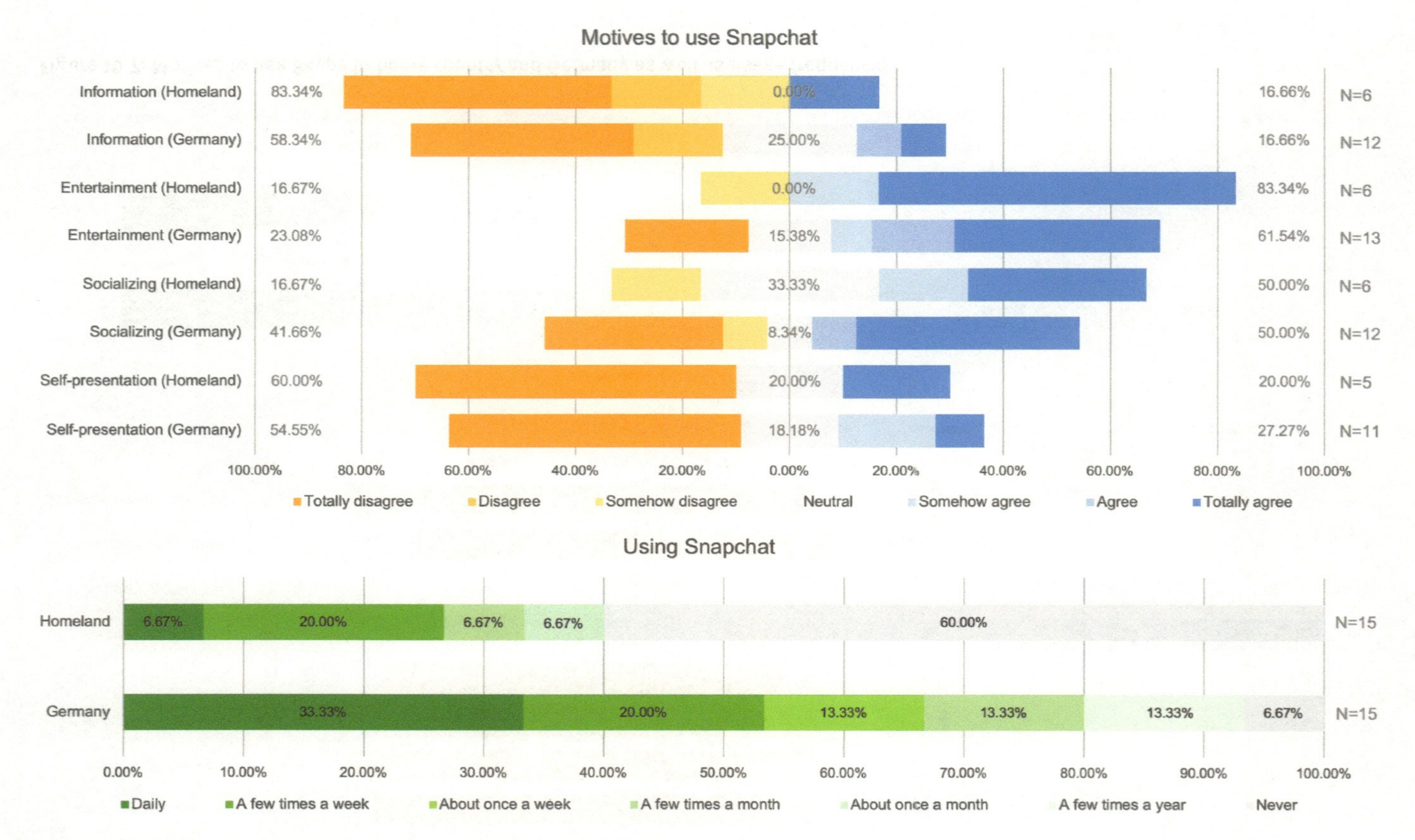

Figure 10.8: Motives to use Snapchat in home country and Germany as well as usage frequency.

Snapchat

According to the data shown in Figure 10.8, the primary motive to use Snapchat is entertainment. Nearly 85 percent (83.34%) of the asylum seekers agreed that they have used Snapchat in their homeland to get entertained. The results show a slight decrease for entertainment as a motive in Germany (61.54%). Another important aspect to use Snapchat is socializing. Half of the asylum seekers agreed they were motivated to use Snapchat in Germany in order to socialize. Snapchat is not a service where refugees and asylum seekers try to receive information. Most asylees disagreed to use Snapchat for information (homeland: 83.34% disagreed; Germany: 58.34% disagreed) or self-presenting purposes (homeland: 60.00% disagreed; Germany: 54.55% disagreed).

Many asylum seekers started to use Snapchat in Germany, while 60 percent (N=15) of those who indicated they use Snapchat did not use it in their homeland. Only 6.67% stated they have used Snapchat on a daily basis and 20 percent indicated they have used Snapchat a few times a week in the home country. The usage frequency of Snapchat in Germany has increased as well. One third (33.33%; N=15) uses Snapchat daily and an additional 20 percent use Snapchat a few times a week. Total 80 percent use Snapchat at least a few times a month (13.33%) or even more frequently in Germany.

LinkedIn

A total of 11 asylum seekers stated they have used LinkedIn in their home country or they use LinkedIn in Germany (Table 10.2). For the usage frequency in the homeland, 40.00% (N=10) of the participants who answered to have used or use LinkedIn indicated they have never used it in their homeland (Figure 10.9). It was stated that 20.00% each have used it a few times a week, a few times a month, or a few times a year. For the situation in Germany, 20% each (N=10) indicated to use LinkedIn daily, or a few times a week, and thirty percent each stated to use it a few times a month or a few times a year. Therefore, more asylum seekers use LinkedIn in Germany.

LinkedIn was mainly used for socializing in the homeland (66.67% agreed). With 40.00%, the same amount disagreed and also agreed to use LinkedIn for socializing in Germany. Some asylum seekers agreed they use LinkedIn for information in Germany (40.00%) and also to have used LinkedIn for information in their homeland (40.00%). Neither in the home country (80.00% disagreed) nor in Germany (80.00% disagreed) was or is LinkedIn used for entertainment.

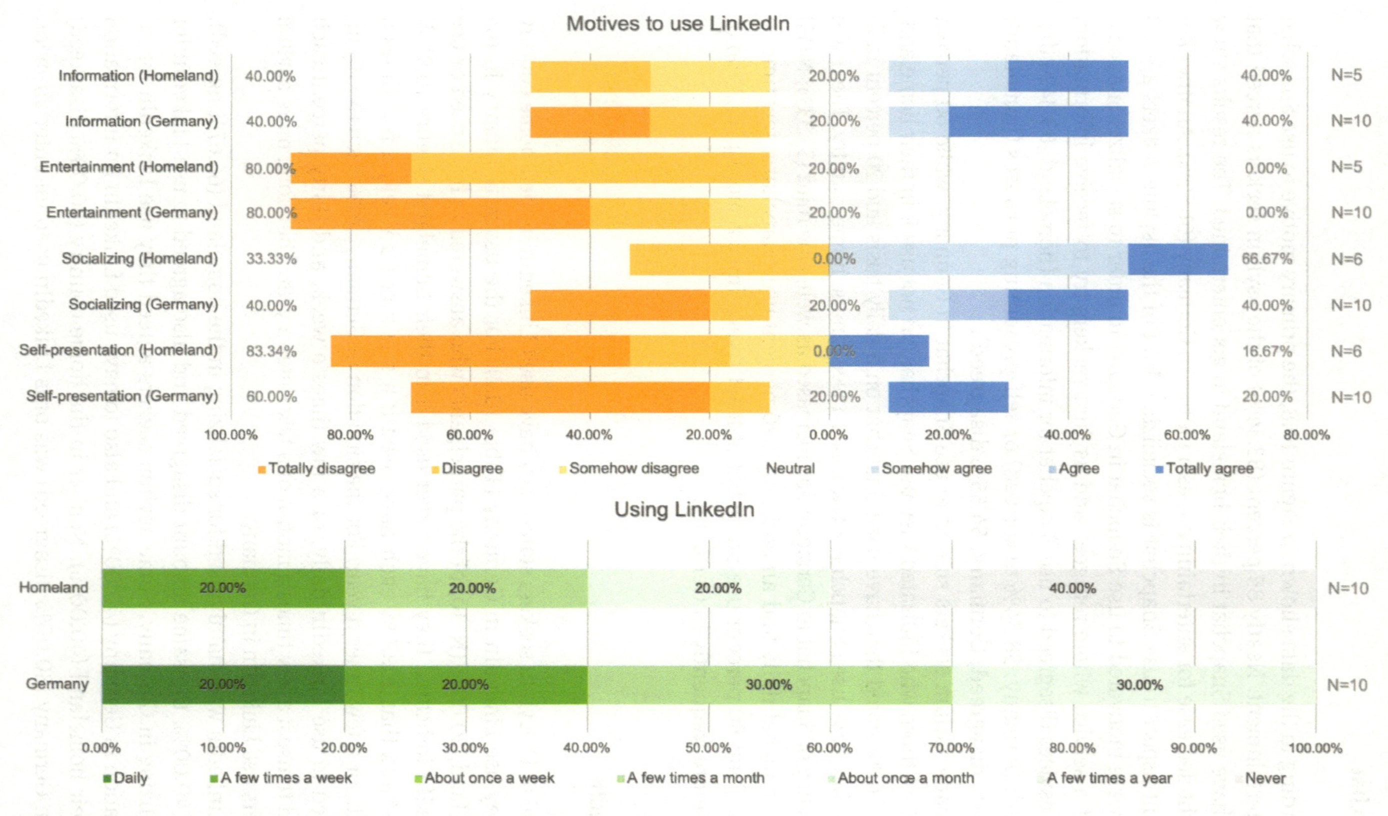

Figure 10.9: Motives to use LinkedIn in home country and Germany as well as usage frequency.

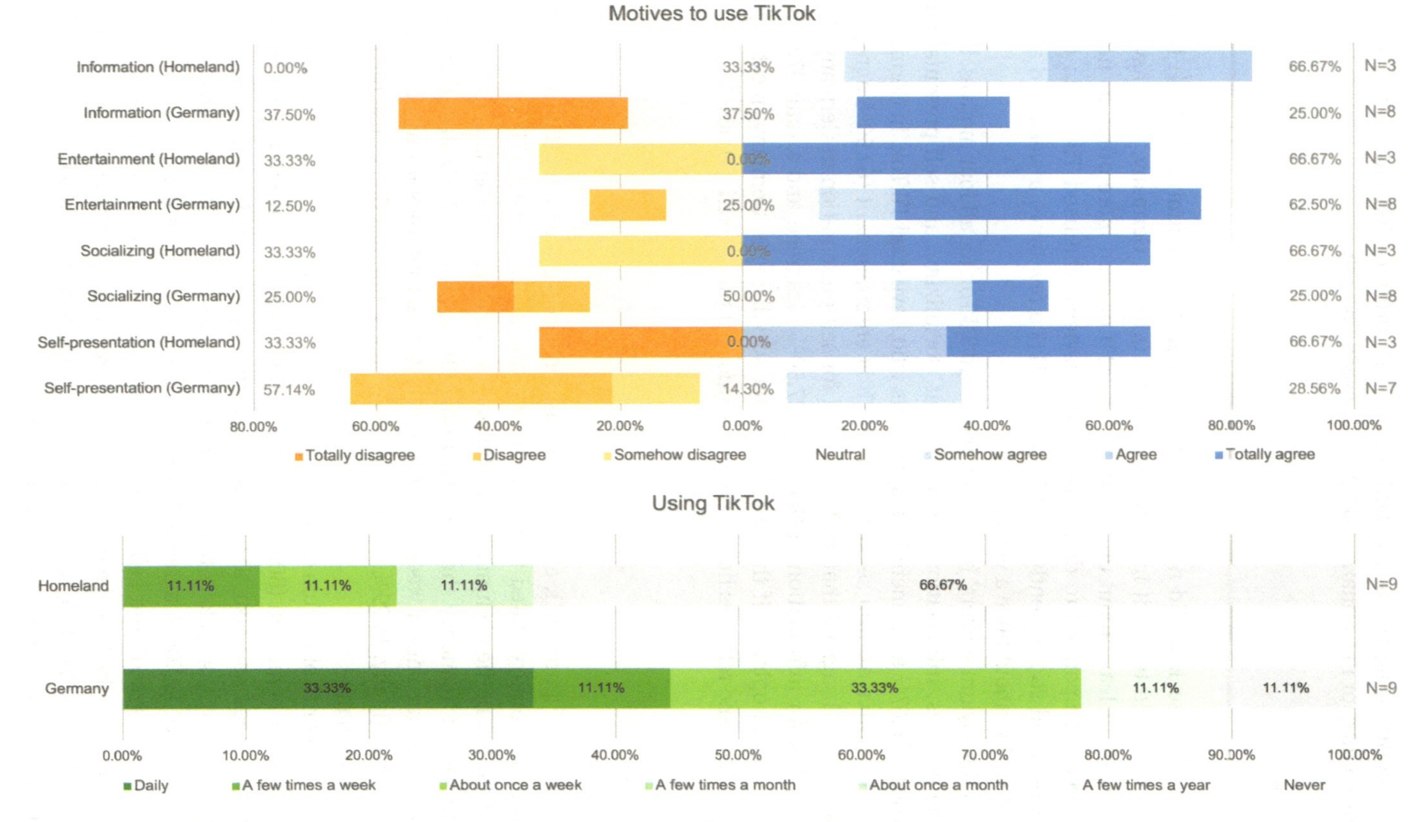

Figure 10.10: Motives to use TikTok in home country and Germany as well as usage frequency.

The majority disagreed to use LinkedIn for self-presentation purposes in their home country or in Germany, 80.00% disagreed on each.

TikTok

Most asylum seekers did not use TikTok in their home country (66.67%; N=9) and just started to use it in Germany (Figure 10.10). In Germany, most asylees use TikTok at least about once a week (33.33%) or even more often, as nearly 80 percent responded to do so (N=9). The asylees who used TikTok in their homeland did so frequently as well, total 11.11% a few times a week and additional 11.11% about once a week.

In the asylum seekers' homeland, they used TikTok for all four motives. Each of the motives, information, entertainment, socializing, and self-presentation reached a 66.67% agreement by respondents who have used TikTok. In Germany, the only motive asylum seekers agree with is entertainment (62.50% agreed). Total 57.14% disagreed that TikTok is used for self-presentation in Germany. To use TikTok for information in Germany, 37.50% disagreed and a total 37.50% had a neutral opinion. If they use TikTok for socializing in Germany, most asylum seekers and refugees have a neutral opinion on this as well (50.00%).

Pinterest

As shown in Figure 10.11, every asylum seeker or refugee who used Pinterest in their home country stated they have used it about once a week (71.43%; N=7). Total 28.57% of the asylum seekers did not use Pinterest. In Germany, the data for the usage frequency show similar results. Here, 71.43% (N=7) state they use Pinterest about once a week, an additional 14.29% (one person) use it once a month. And finally, 14.29% of those who have used Pinterest do not use the platform in Germany.

The majority of asylum seekers agreed they have used Pinterest for entertainment in their homeland (80.00%). If information was a motive to use Pinterest, 40.00% agreed and also 40.00% had a neutral opinion on it. In Germany, the motives for asylum seekers to use Pinterest are information (66.66% agreed) and entertainment (50.00% agreed). Most asylum seekers disagreed that socializing (50.00% disagreed) and self-presentation (75.00% disagreed) were motives to use Pinterest in the homeland. Similar results are shown for the usage of Pinterest in Germany as most disagreed to use Pinterest for socializing (60.00% disagreed) or self-presentation (60.00% disagreed) there.

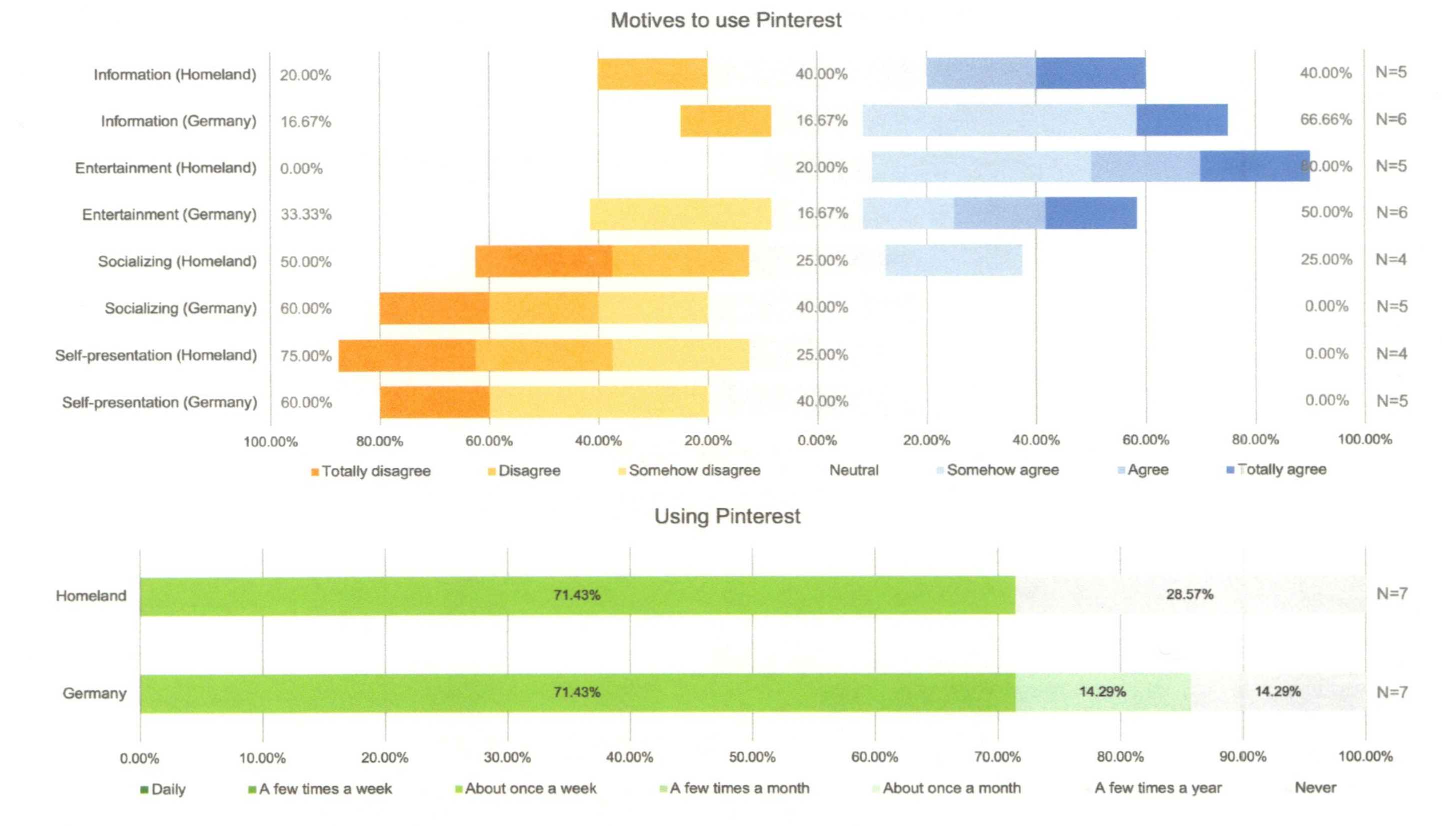

Figure 10.11: Motives to use Pinterest in home country and Germany as well as usage frequency.

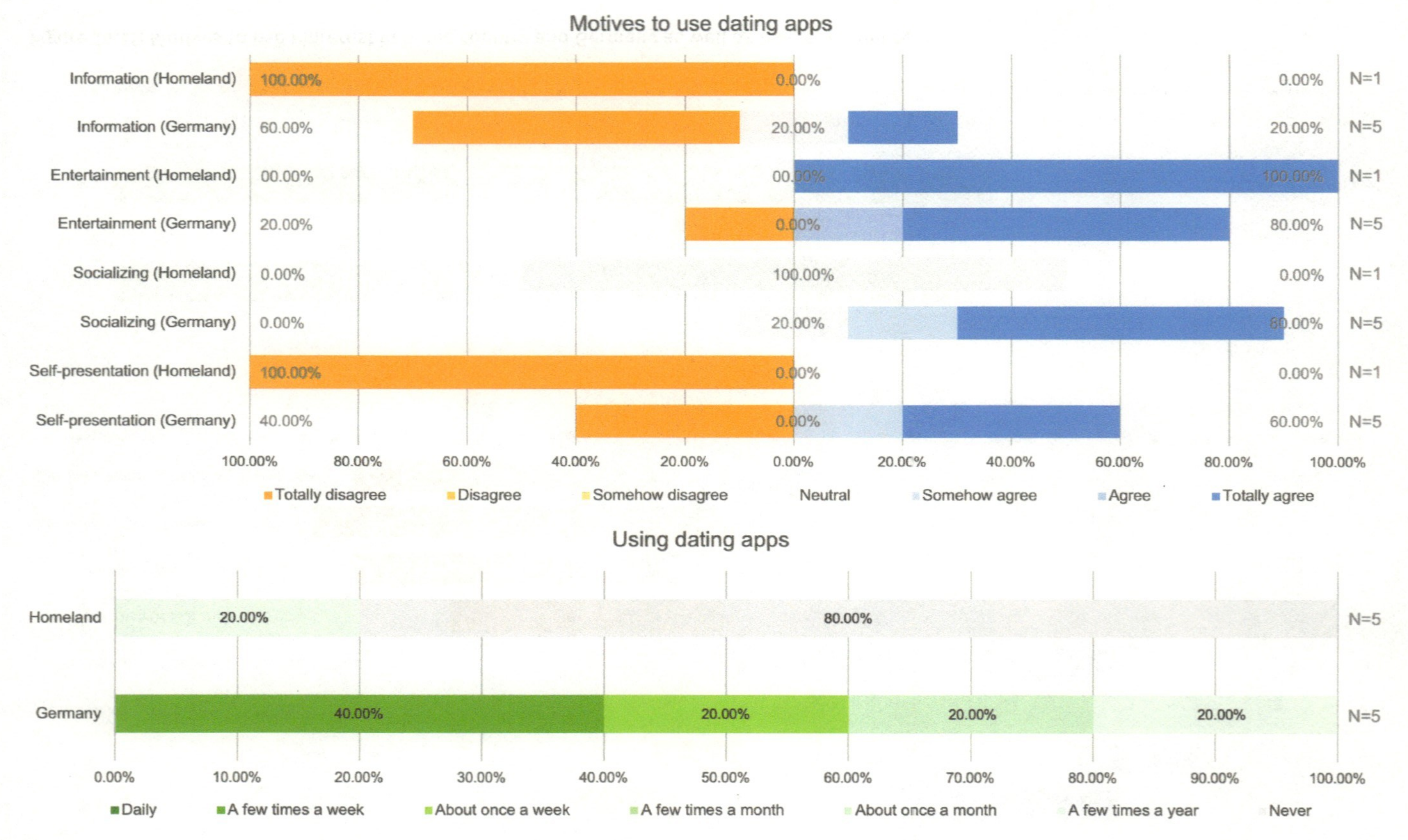

Figure 10.12: Motives to use dating apps in home country and Germany as well as usage frequency.

Dating apps

Dating apps became more important for asylum seekers after arriving in Germany. Only one asylum seeker (20.00%; N=5) indicated he or she used dating apps in the homeland, and he or she did it only a few times a year (Figure 10.12). Therefore, 80.00% indicated they have not used dating apps but started to do so in Germany. Participants use them at least a few times a year in Germany. Total 40.00% (N=5) use dating apps a few times a week and an additional 20.00% use it about once a week.

Motives to use dating apps in Germany are mainly entertainment (80.00% agreed) and socializing (80.00% agreed as well), followed by self-presentation (60.00% agreed). Entertainment (100% agreed) was also a motive for the person who used dating apps in the home country. Information is not a motive to use dating apps at all.

10.2 Online media

E-mail

More asylum seekers began to write e-mails in Germany compared to their home country. They also write e-mails more frequently in Germany (Figure 10.13). Total 22.81% (N=57) who answered they write e-mails did not do it in their homeland. But more than 50 percent answered that they have written e-mails in their homeland at least about once a week (15.79%) or more frequently. An amount of 17.54% has written e-mails daily in the homeland. For Germany, nearly 90.00% (N=58) stated they write e-mails about once a week (13.79%) or even more often. A total of 44.83% write e-mails on a daily basis in Germany.

Most of the respondents agreed that they write e-mails for information purposes (73.08% agreed) in Germany. While living in their home country, 65.00% agreed that they have written e-mails to receive or share information. Looking at entertainment as a motive to write e-mails, both, in the homeland and in Germany alike, around 50.00% disagreed and around 25 percent had a neutral opinion on it. For socializing as a motive, 54.00% agreed for Germany and 60.00% for the home country. Self-presentation is a motive for them to write e-mails in Germany (40.00% agreed), whereas only 37.15% agreed that they have written e-mails for self-presentation purposes in their homeland.

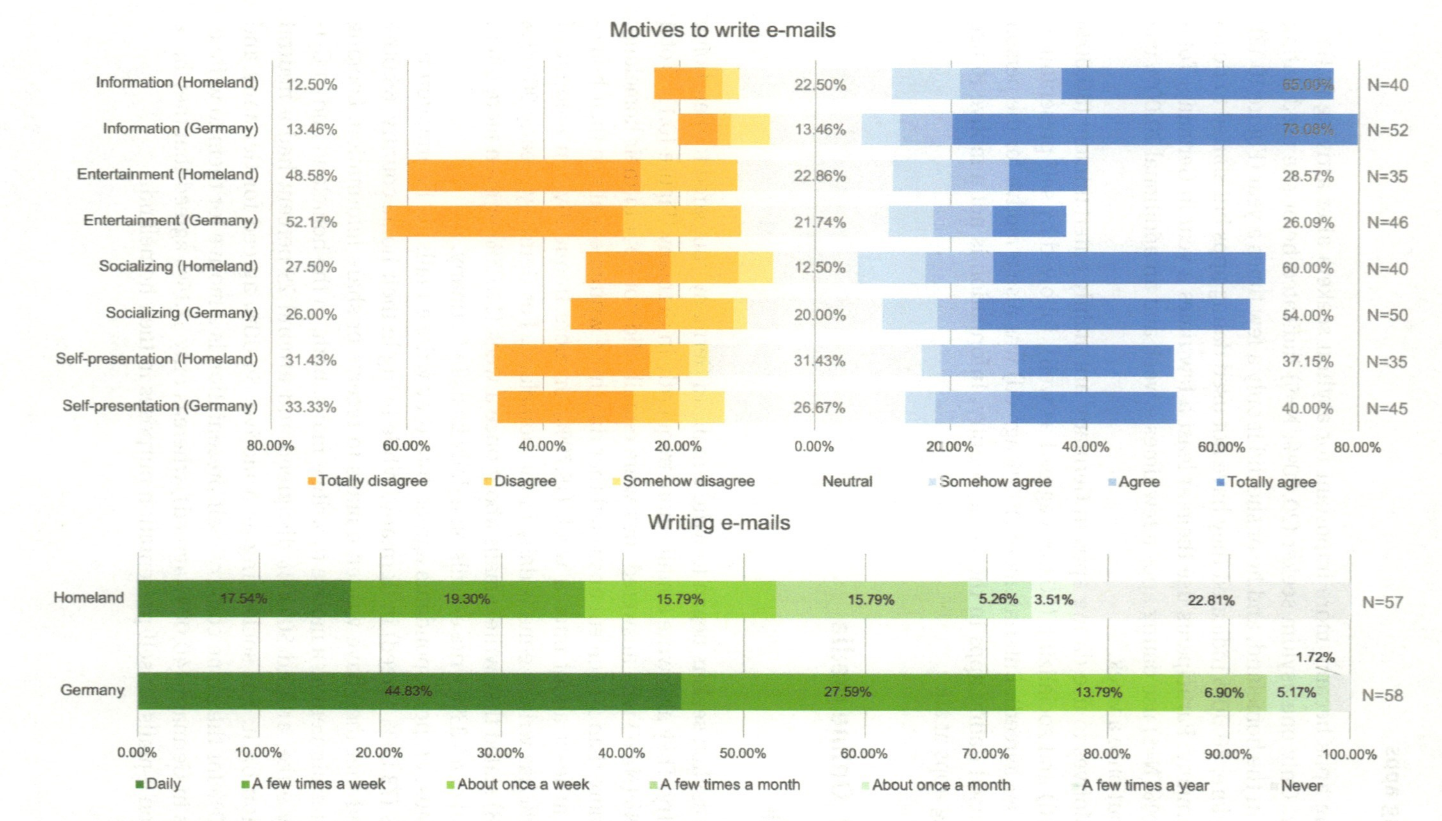

Figure 10.13: Motives to use e-mail in home country and Germany as well as usage frequency.

Search engines

Search engines are used more often by asylum seekers since living in Germany (Figure 10.14). An amount of 91.11% (N=45) stated to use search engines on a daily basis in Germany, an additional 4.44% indicated to use search engines a few times a week. Further 2.22% use search engines about once a week and another 2.22% use search engines a few times a month in Germany. In the homeland, only 69.57% (N=46) have used search engines daily and 15.22% a few times a week. Total 10.87% have used search engines about once a week and an additional 4.35% a few times a month.

Translation services

The services Google Translate and DeepL are examples for translation services. Taking a look at the usage frequency of translation services and comparing the use in the homeland and in Germany, many asylum seekers began to use translation services in Germany (Figure 10.15). Nearly 30 percent (N=53) stated they have never used translation services in their home county. Whereby only 24.53% have used translation services on a daily basis, nearly 90 percent (N=54) of the asylum seekers indicated to use translation services daily in Germany. Additional 9.26% stated to use them a few times a week, and 1.85% use these services about once a week. Accordingly, the translation services are used much more frequently by asylum seekers in Germany.

Encyclopedia

The majority of asylum seekers use encyclopedias in order to receive information (Figure 10.16). Around 88.45% agreed they used encyclopedias for information and knowledge in the homeland and around 84.61% use them for information in Germany. Another motive that many asylum seekers agree on is socializing, meaning that in order to be able to communicate about certain topics with other people, they want to obtain knowledge by reading encyclopedias. About 61.53% agreed in both cases, for the situation back in the homeland and for the situation in Germany (57.69%). An amount of 44.00% also agreed they have used encyclopedias for self-presentation purposes in their home country, but the same amount disagreed (44.00%) as well. With a decrease of 16 percent points, only 28.00% agreed and 48.00% disagreed to be motivated to use encyclopedias because of self-presentation purposes in Germany. For the entertainment motive, the same

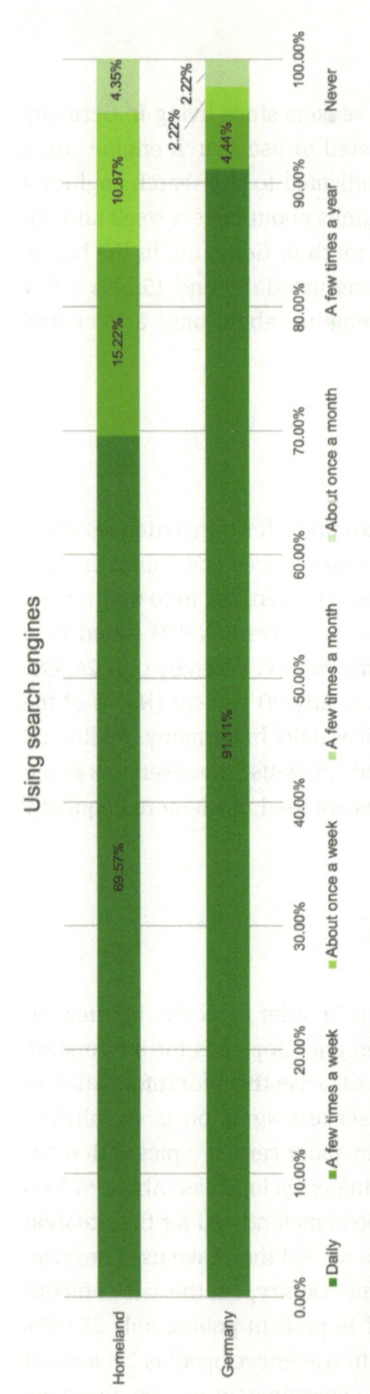

Figure 10.14: Usage frequency of search engines in home country and Germany.

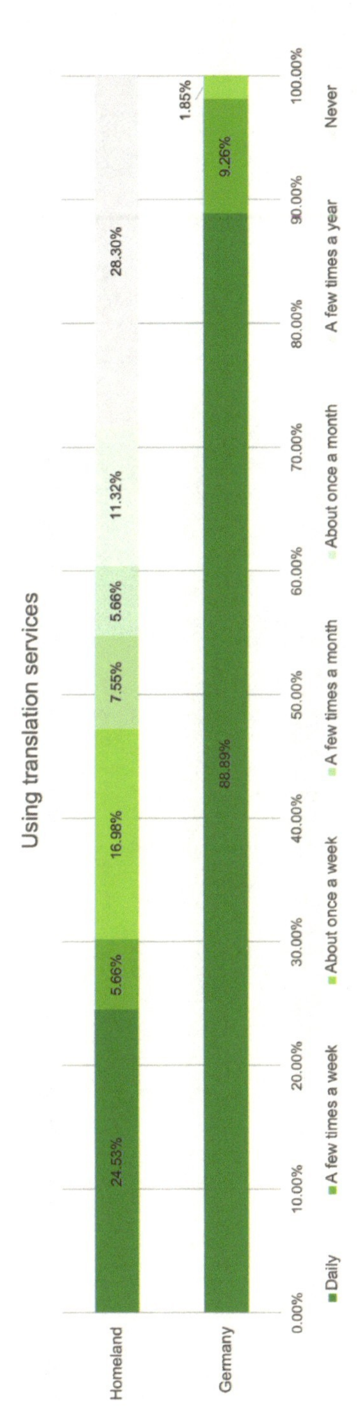

Figure 10.15: Usage frequency of translation services in home country and Germany.

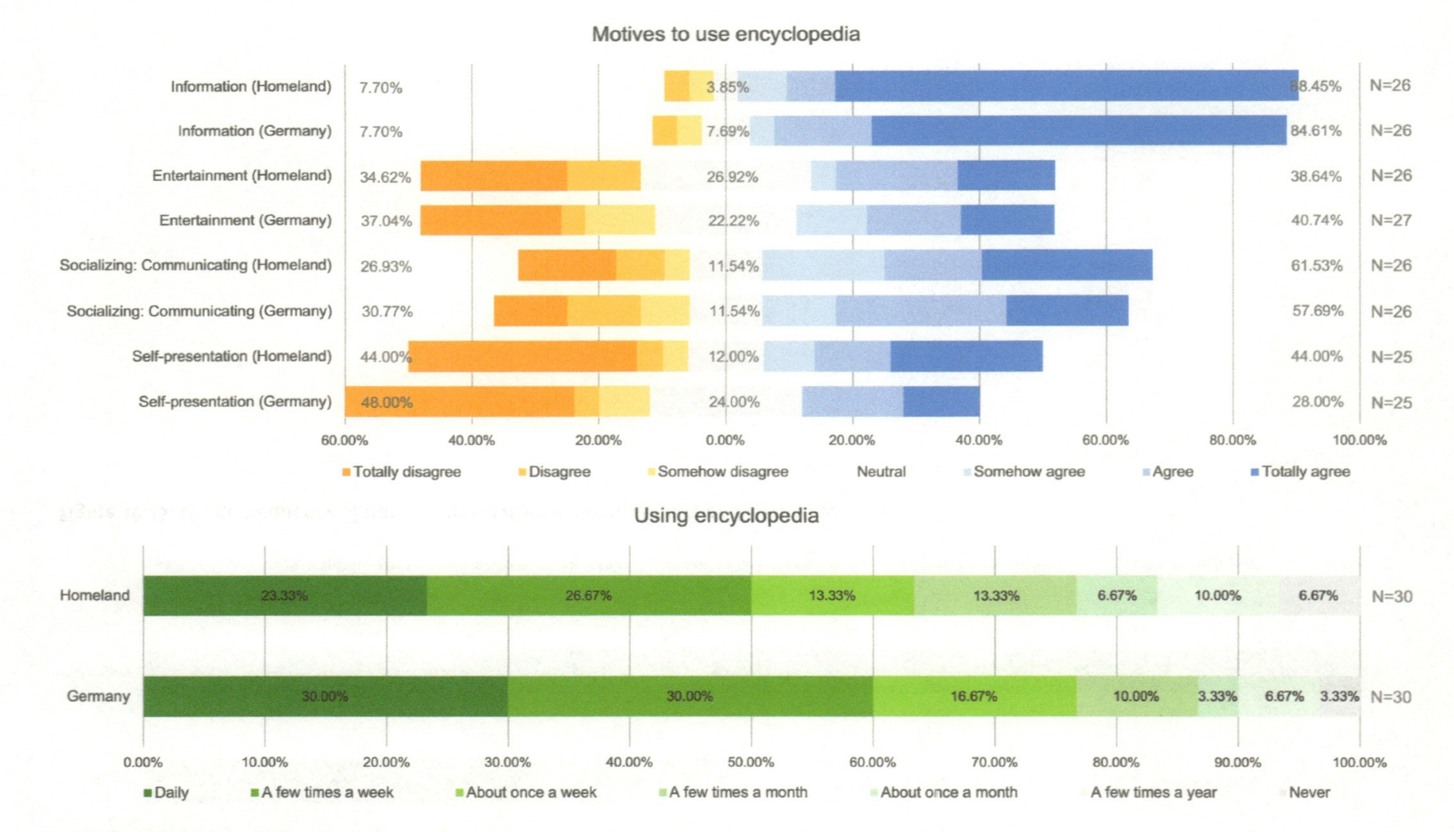

Figure 10.16: Motives to use encyclopedias in home country and Germany as well as usage frequency.

amount of around 35 to 40 percent agreed and disagreed for the homeland (38.64% agreed; 34.62% disagreed) and for Germany (40.74% agreed; 37.04% disagreed).

Encyclopedias are used more frequently in Germany. Here, 30.00% (N=30) use encyclopedias daily, an additional 30.00% use these services a few times a week, and 16.67% use them about once a week. In the homeland, 23.33% (N=30) have used encyclopedias daily, 26.67% a few times a week, and 13.33% have used them about once a week. Total 6.67% stated they did not use encyclopedias in their home country and 3.33% indicated they do not use them in Germany.

Streaming services

For streaming services (e.g., Netflix or Amazon Prime) a clear shift in the usage behavior can be observed (Figure 10.17). While 58.06% (N=31) stated they have not used streaming services in their home country, all survey respondents who indicated to use streaming services do so in Germany (100%; N=31). In the homeland, 12.90% have used streaming services on a daily basis and an additional 12.90% have used them a few times a week, whereas in Germany 38.71% use streaming services on a daily basis and 29.03% use them a few times a week. Therefore, more asylum seekers use streaming services in Germany and further, it is noticeably that they use these services more frequently.

The most important motive for most asylum seekers to use streaming services is entertainment. In the home country, with an amount of 83.33%, most asylum seekers agreed they have used streaming services in order to get entertained. For Germany, 62.50% agreed they are motivated by entertainment. While 41.67% agreed they have used streaming services for information, a total 50.00% stated to use them in Germany to this end. For socializing and therefore in order to be able to communicate with others, the same amount agreed they were motivated by it in the home country (45.45%) and in Germany (45.83%) as well. To build their personal identity was a motive to use streaming services for 54.54% in the homeland. In Germany, around 41.67% agreed to use streaming services in order to build their personal identity.

Learning apps

According to the data presented in Figure 10.18, more asylum seekers started to use learning apps (e.g., language learning apps such as Duolingo) in Germany. While 44.44% (N=27) have never used learning apps in their home country,

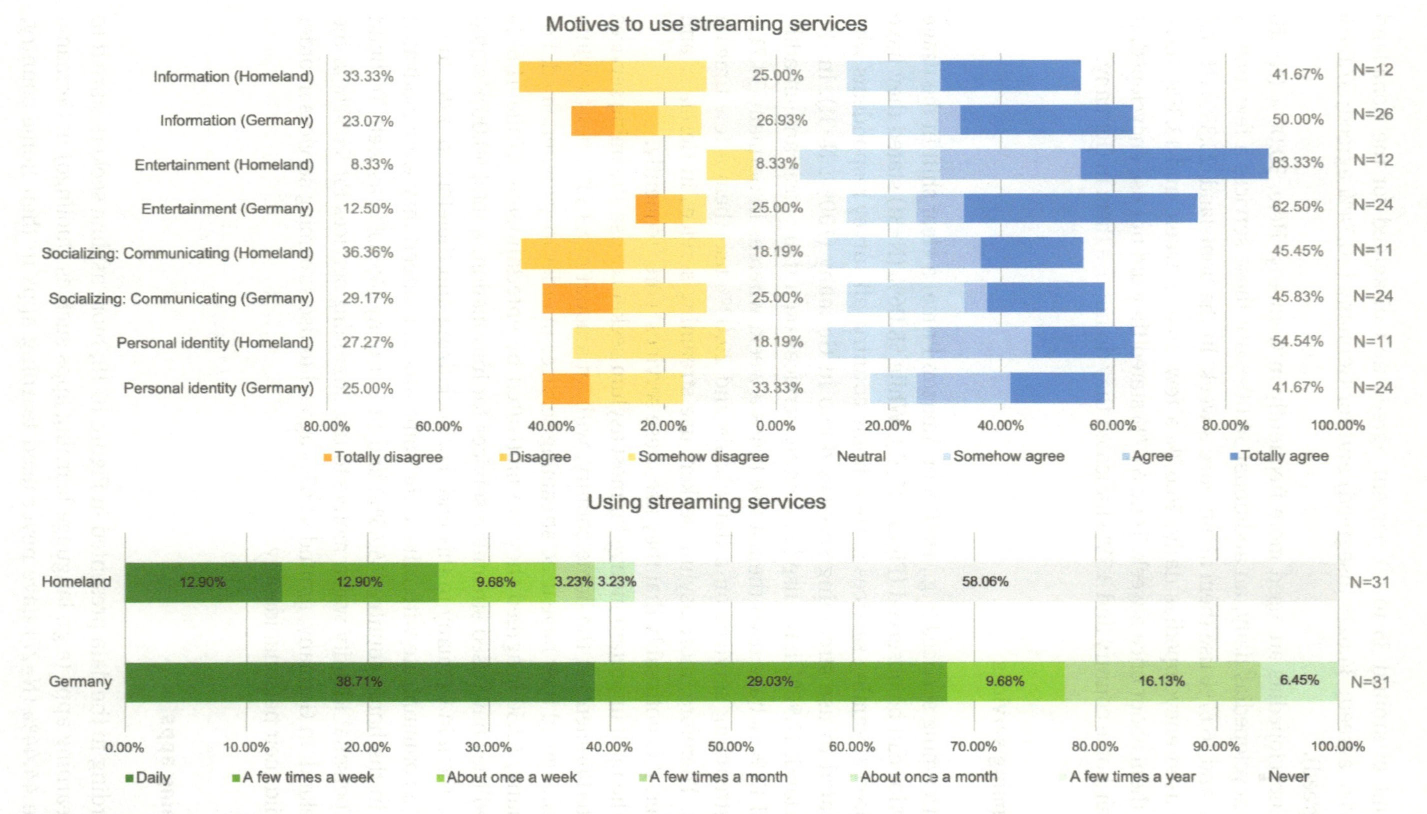

Figure 10.17: Motives to use streaming services in home country and Germany as well as usage frequency.

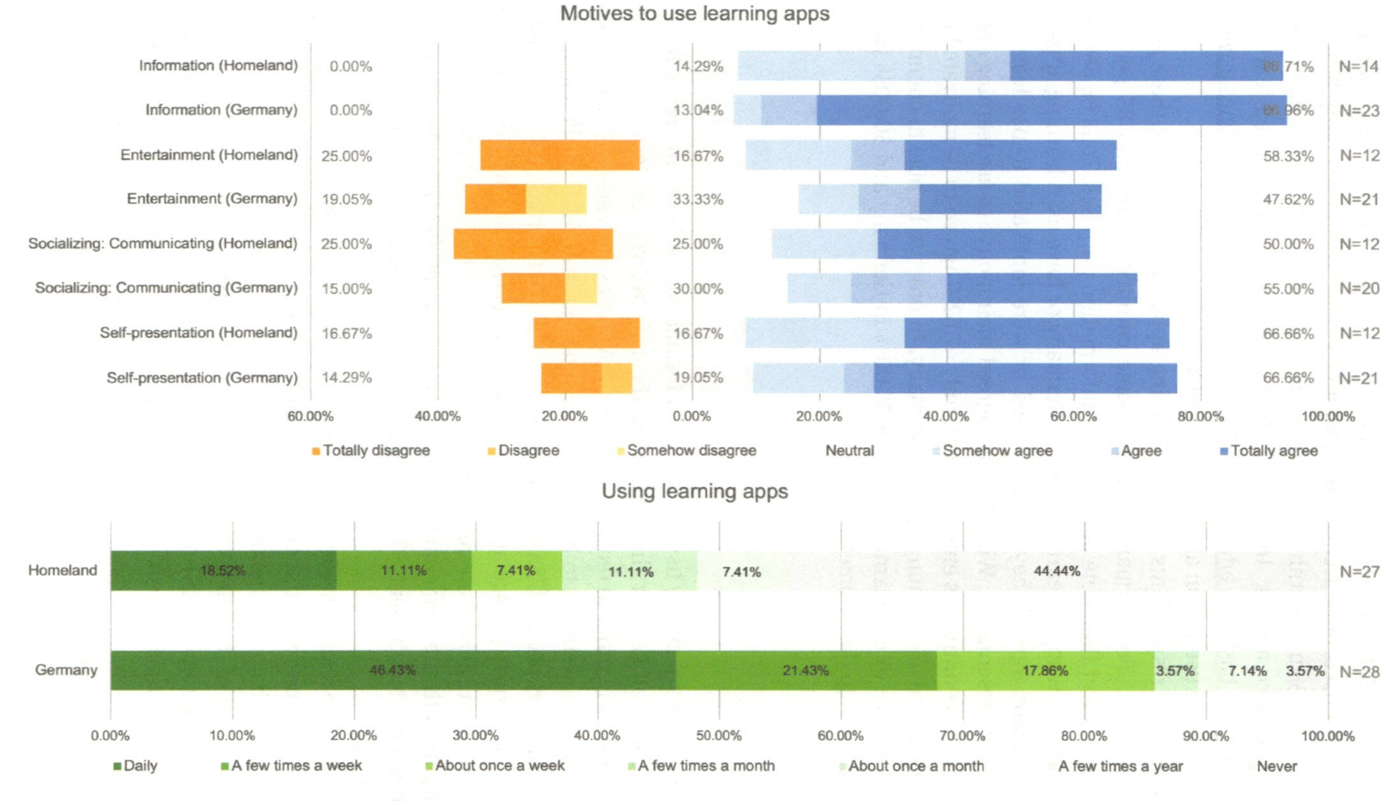

Figure 10.18: Motives to use learning apps in home country and Germany as well as usage frequency.

only 3.57% (N=28) stated they never use learning apps in Germany. Learning apps are also used more frequently now, 46.43% use them on a daily basis in Germany, 21.43% use learning apps a few times a week, and 17.86% have used them about once a week. In the homeland, only 18.52% have used learning apps on a daily basis, an additional 11.11% have used them a few times a week and 7.41% have used them about once a week.

Most of the respondents who use learning apps agreed they use these apps for information. For Germany and the home country, more than 85 percent agreed they are or they were motivated by information purposes. Total 66.66% of asylum seekers who use learning apps in Germany or have used learning apps in their homeland agreed they are motivated or were motivated to use learning apps for self-presentation. While 58.33% agreed their motive to use learning apps in their home country was entertainment, only 47.62% agreed to be motivated by entertainment to use learning apps in Germany. In order to be able to communicate is a motive to use learning apps for 55.00% in Germany and 50.00% were motivated by it in their homeland.

News websites

News websites are mostly used to receive information, as shown in Figure 10.19. Total 85.00% agreed for the home country and 78.26% for Germany to use news websites in order to receive information. Many asylum seekers also agreed they used news websites to be able to socialize and therefore to talk about the news with other people (75.00%); only 52.17% agreed their motive to use news websites in Germany is socializing, resulting in a decrease of around 20 percent points. Asylum seekers are motivated to use news websites for their personal identity in both the homeland (60.00%) and Germany (52.17%). To use news websites for entertainment, 52.63% agreed for the homeland and 45.00% agreed for Germany.

A few more asylum seekers use news websites in Germany, as 8.00% (N=25) indicated they never used them in their home country. The data shows no major differences in the usage frequency of news websites by asylum seekers comparing their usage in the homeland and in Germany. News websites are used on a daily basis (60.00%, N=25) in Germany and an additional 28.00% indicated to use them a few times a week. In the homeland, 64.00% have used news websites on a daily basis and four percent stated they have used them about once a week in the home country.

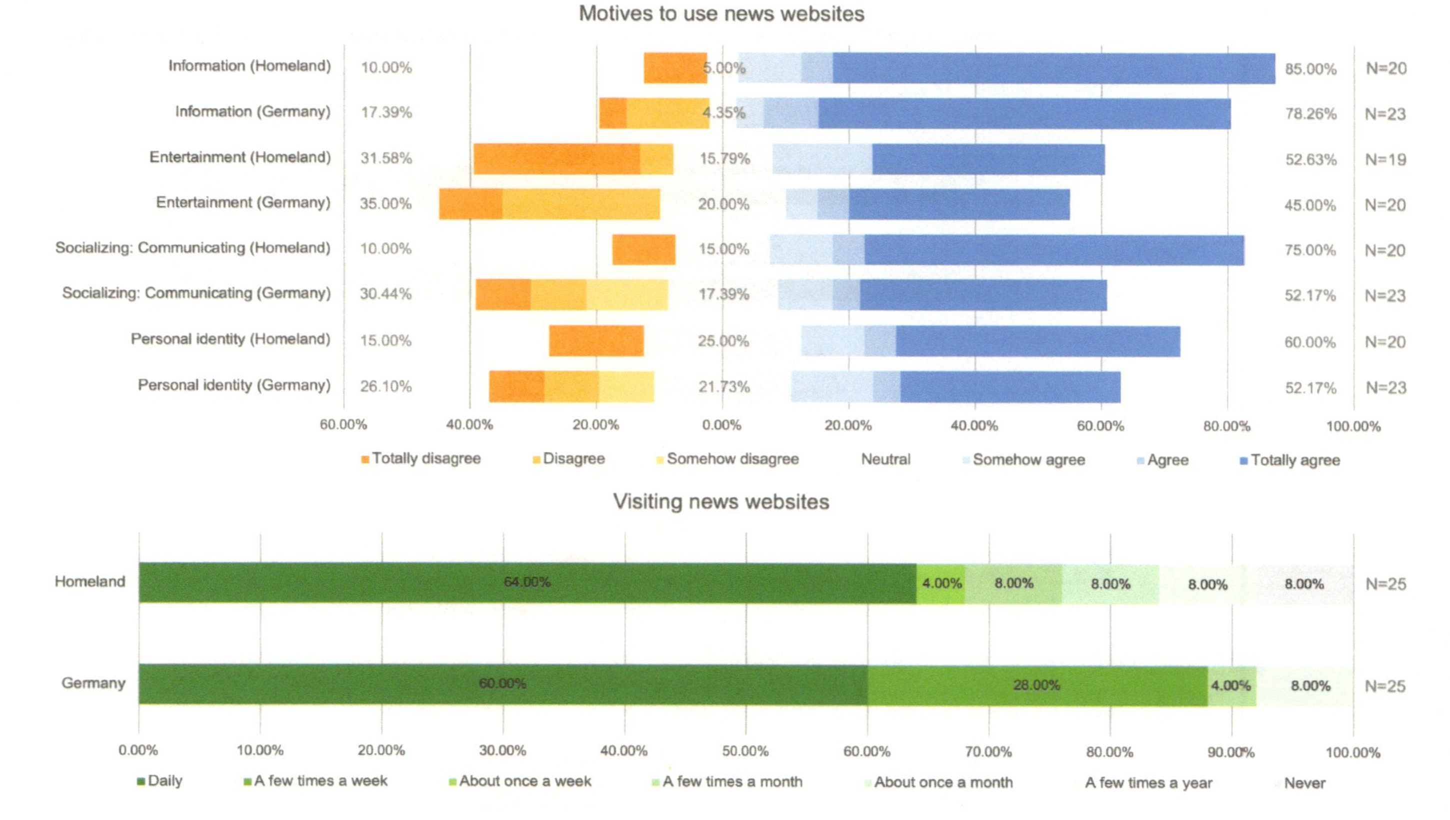

Figure 10.19: Motives to use news websites in home country and Germany as well as usage frequency.

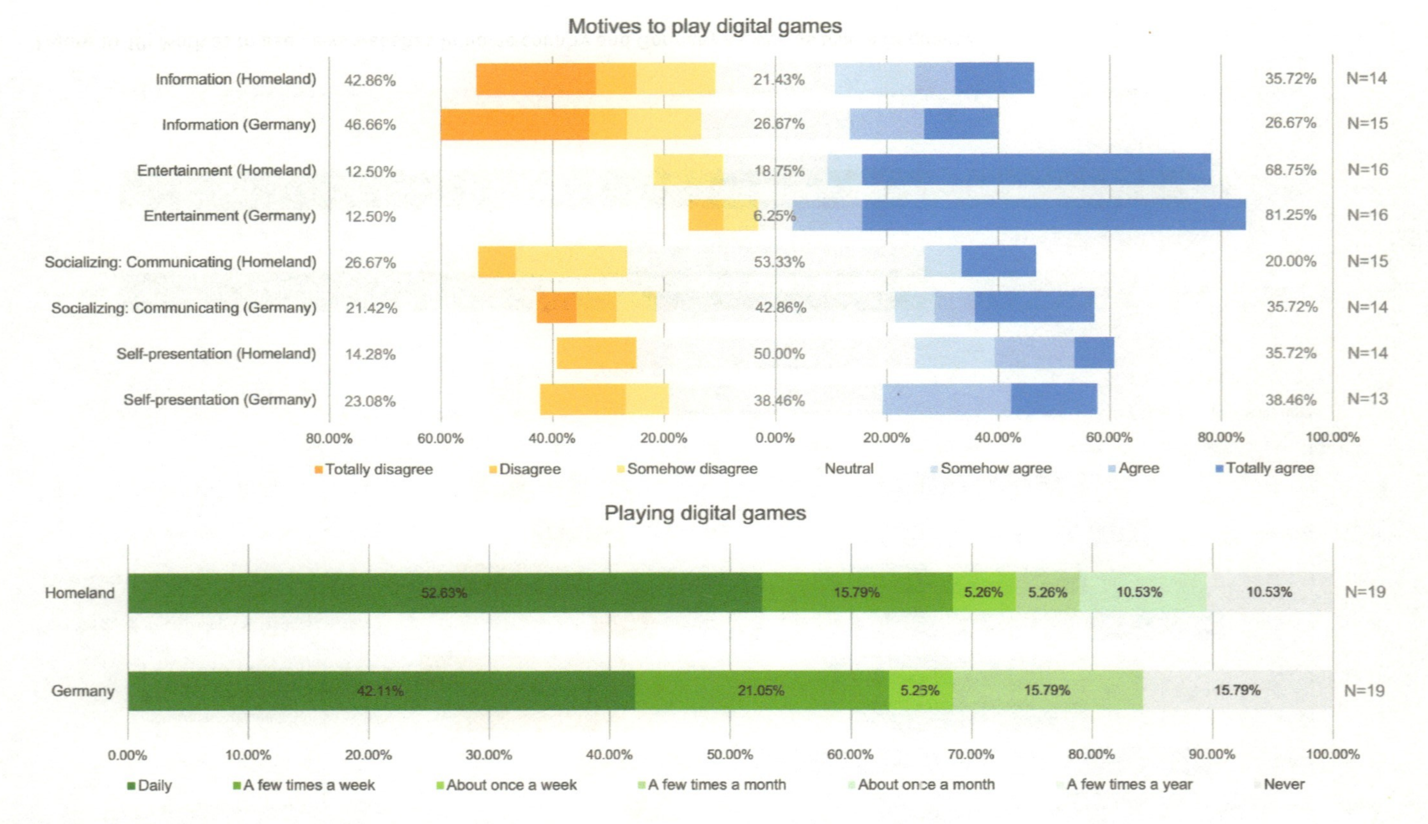

Figure 10.20: Motives to use digital games in home country and Germany as well as usage frequency.

Digital games

Entertainment is by far the most important reason why asylum seekers have played digital games in their home country (68.75%) and play digital games in Germany (81.25%) (Figure 10.20). For the situation in the home country, half (50.00%) of the respondents had a neutral opinion on if they play digital games because of self-presenting purposes and total 35.72% agreed. For the situation in Germany, 38.46% had a neutral opinion, and 38.46% also agreed. Playing digital games because of information is not a motive for asylum seekers as around 45 percent disagreed for both situations, in the home country (42.86% disagreed) and in Germany (46.66% disagreed). Around 50 percent (53.33%) have a neutral opinion on whether they played digital games in their home country to be able to communicate (socializing). In Germany, most asylum seekers also have a neutral opinion on this (44.86%).

The amount of asylum seekers who play digital games as well as the playing frequency decreased in Germany. While 52.63% (N=19) stated they have played digital games on a daily basis, 15.79% played digital games a few times a week, and additional 5.26% about once a week in the home country, total 42.11% (N=19) indicated they play digital games daily, 21.05% indicated they play them a few times a week, and additional 5.26% about once a week in Germany. In the home country, 10.53% (N=19) have not played digital games, and in Germany, 15.79% do not play digital games.

Podcasts

Total 50.00% (N=10) of asylum seekers who listen to podcasts stated to do so a few times a week and an additional 10.00% even listen to podcasts daily in Germany (Figure 10.21). For the homeland, 40.00% (N=10) stated they did not listen to podcasts. Total 20.00% listened to podcasts a few times a week in the home country and an additional 10.00% did so a few times a week. Consequently, more asylum seekers are listening to podcasts in Germany and they do it more frequently.

Asylum seekers agreed to listen to podcasts to receive information in Germany; here, total 90.00% of the respondents agreed. Another important motive why asylum seekers listen to podcasts in Germany is to build their personal identity; here, 70.00% agreed. Half of the ones who listen to podcasts in Germany also agreed that they are motivated because of socializing, meaning in order to be able to talk with others about the podcasts. In the homeland, the most important motives to listen to podcasts were information (50.00% agreed)

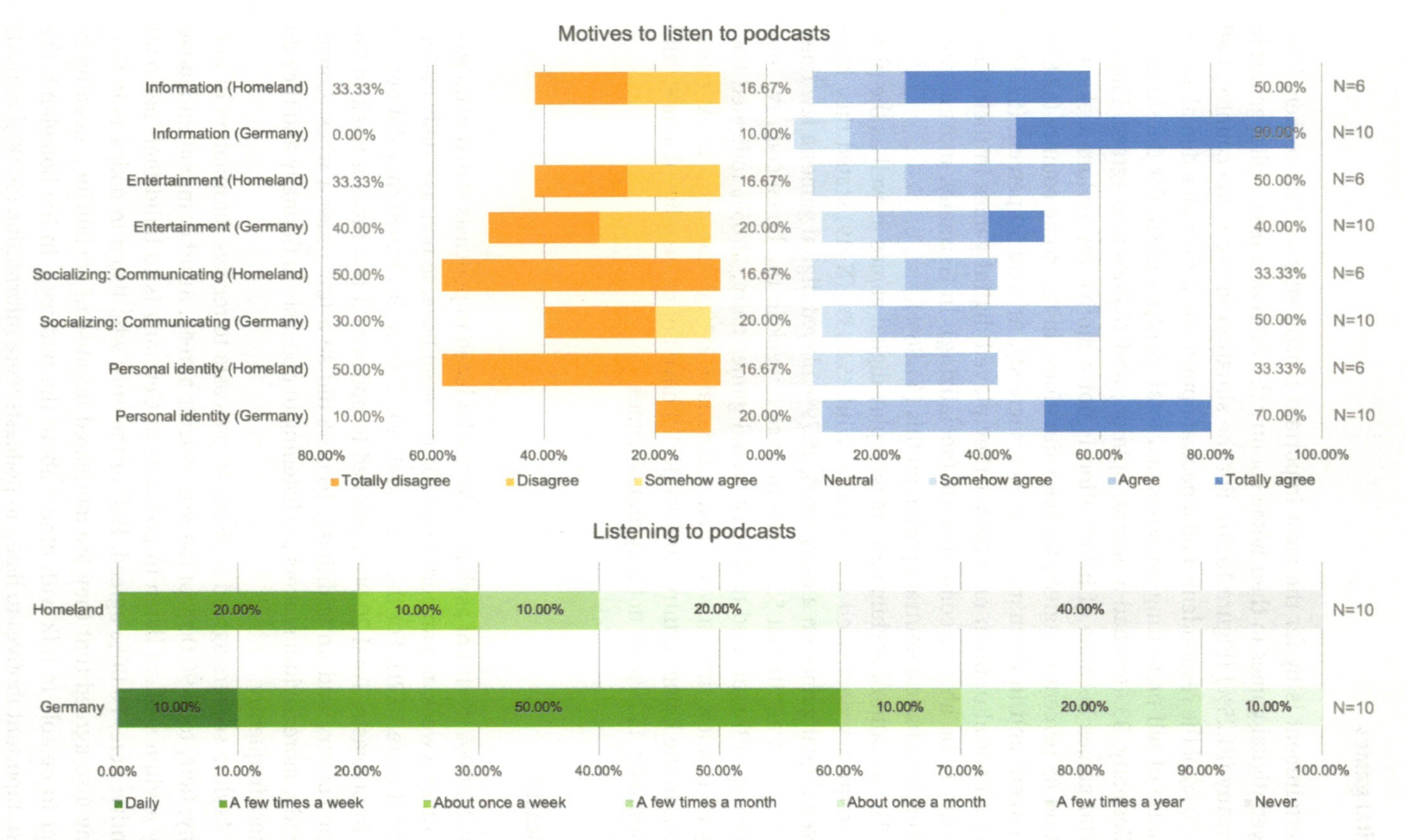

Figure 10.21: Motives to use podcasts in home country and Germany as well as usage frequency.

and entertainment (50.00% agreed). For socializing and personal identity, even more people disagreed they were motivated by it in the homeland (50.00% disagreed on each).

10.3 ICT

Smartphone

The smartphone is the most frequent as well as most used ICT device. According to Figure 10.22, all asylum seekers who indicated to use smartphones, use them on a daily basis in Germany (100%; N=52). Also, 98.04% stated they have used the smartphone daily in their home country and an additional 1.96% (N=51) stated they have used the smartphone a few times a week.

Many asylum seekers agreed to be motivated to use smartphones because of all four motives, information, entertainment, socializing, and self-presentation. With 90.00%, most respondents agreed to use the smartphone for information in Germany. Following, 87.76% agreed to be motivated to use the smartphone for socializing purposes in Germany. For the homeland, 86.00% agreed on each of the two motives – information and socializing. Around 76.60% agreed to be motivated to use smartphones for entertainment in Germany and also 77.55% agreed they were motivated to use smartphones for entertainment in the homeland. Self-presentation is the only motive that smartphones were used more often for by asylum seekers in the homeland (68.08% agreed to homeland; 62.50% agreed to Germany).

Computer/notebook/laptop

Most asylum seekers agreed to use computers, notebooks, or laptops for information (Figure 10.23). Around 65 percent agreed to both, to have used it for information in the homeland (66.67% agreed) and to use it for information in Germany (69.04% agreed). Entertainment is also a motive for asylum seekers to use a computer, notebook, or laptop. Total 63.42% agreed they have used at least one of the devices for entertainment in the home country and a total of 60.00% agreed to use it for entertainment in Germany. The computer, notebook, or laptop was not used and is not used by most asylum seekers for self-presentation purposes (38.46% agreed for homeland and 38.46% agreed for Germany).

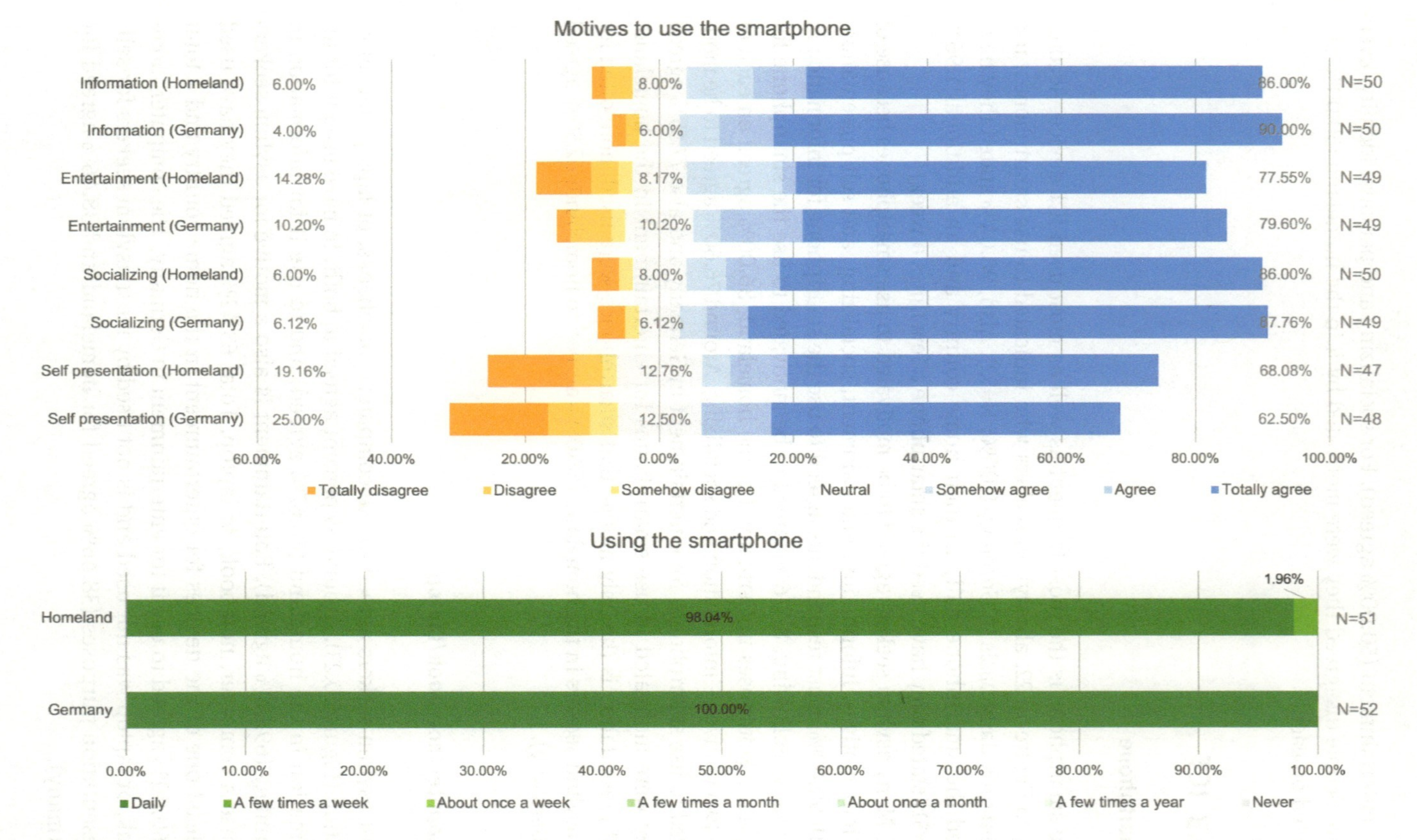

Figure 10.22: Motives to use smartphones in home country and Germany as well as usage frequency.

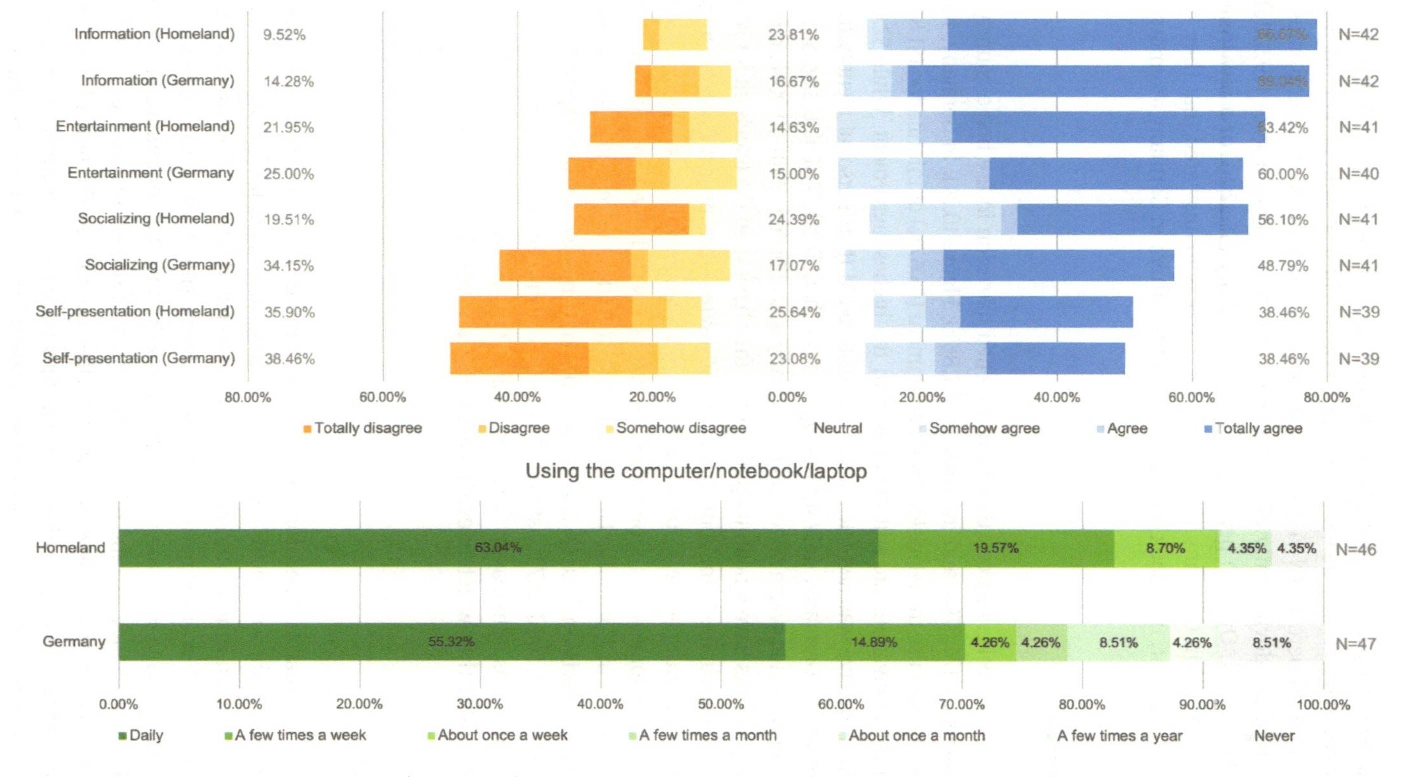

Figure 10.23: Motives to use computers/notebooks/laptops in home country and Germany as well as usage frequency.

The usage frequency of the computer, notebook, or laptop decreased in Germany compared to the usage frequency in the asylum seekers homeland. While more than 80 percent (N=46) used the computer, notebook, or laptop, most of them used the device daily (63.04%), and some a few times a week (19.57%) in their homeland. Only 70 percent (N=47) use it daily (55.32%), or a few times a week (14.89%) in Germany. An amount of 8.51% who answered that they have used a computer, notebook, or laptop do not use them in Germany anymore.

Television

Most asylum seekers agreed they have watched television in the home country for entertainment (74.07%) and information (67.86%; Figure 10.24). In Germany, fewer asylum seekers use the television for entertainment (66.67%) or information (51.85%), but these motivations are still the ones with the highest agreement. While 52.00% agreed they have watched television in order to build their personal identity in the homeland, in Germany, only 42.31% agreed on that. Total 37.50% disagreed that their motivation to watch television in Germany is socializing. With 40.00%, most of the respondents have a neutral opinion on if they have used the television for socializing purposes in the home country and 36.00% agreed they used it for socializing.

Fewer asylum seekers and refugees watch television in Germany when compared to the homeland. In the homeland, 9.09% (N=33) have never used the television, while in Germany, 12.12% (N=33) stated they never use the television. Also, the television is used less frequently in Germany when compared to the home country. Total 72.73% stated they have watched television in the homeland on a daily basis, whereas in Germany, only 39.39% watch it daily. An additional 27.27% stated they watch television a few times a week in Germany.

Tablet

Tablets are mainly used for entertainment by asylum seekers and refugees. Figure 10.25 shows that 66.66% agreed they have used tablets in the homeland for entertainment and 69.24% agreed to use the tablet for entertainment in Germany. For information, the data shows that half of the respondents agreed they have used the tablet for information in the homeland, and a total 46.15% agreed they use the tablet in Germany in order to receive or search for information. The data show that most asylees do not use the tablet for socializing. An amount of 41.67% disagreed that they have used the tablet for socializing in the homeland and 50.00% disagreed to use it for socializing in Germany. With

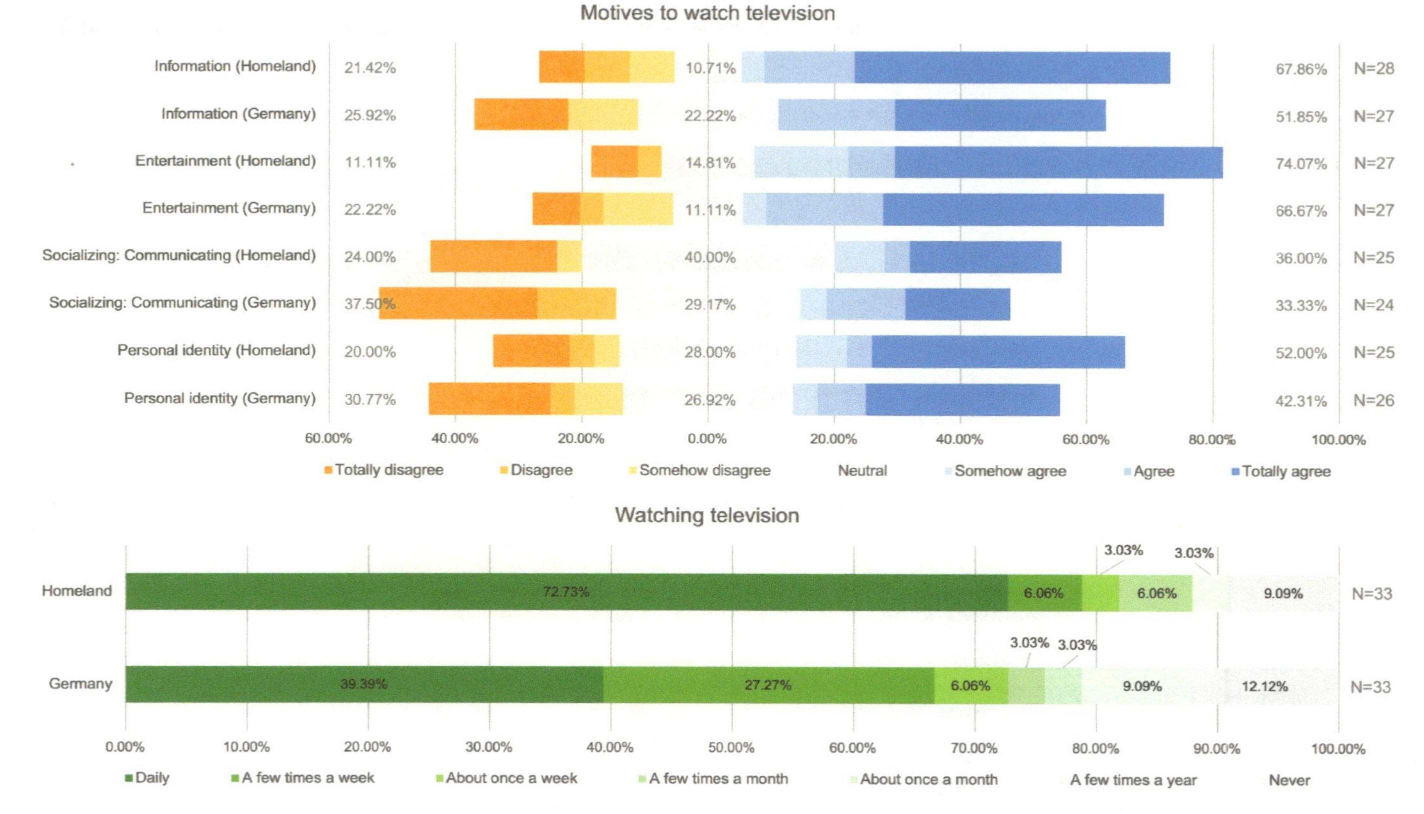

Figure 10.24: Motives to use televisions in home country and Germany as well as usage frequency.

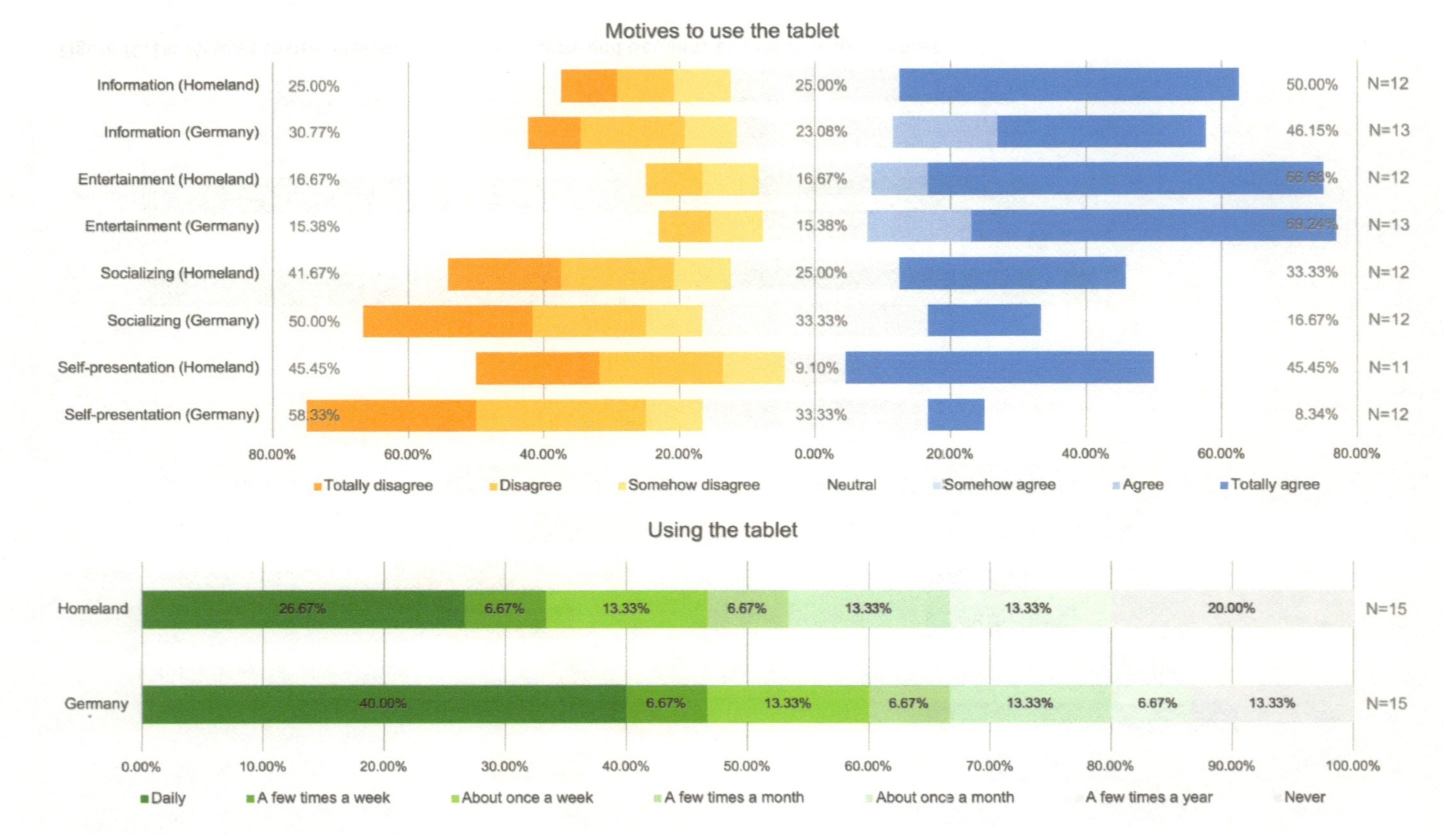

Figure 10.25: Motives to use tablets in home country and Germany as well as usage frequency.

45.45%, the same amount agreed and disagreed that they used the tablet for self-presentation purposes in the homeland. Most (58.33%) disagreed that they use the tablet for self-presentation in Germany.

Total 20.00% (N=15) of the survey attendees who indicated they generally use tablets did not use them in their home country. For Germany, 13.33% (N=15) indicated they do not use tablets. Total 40.00% of the asylees use the tablet on a daily basis, whereby only 26.67% indicated they have used the tablet daily in the homeland. Consequently, more asylum seekers are using the tablet in the target country Germany, and most asylum seekers use the tablet more frequently in Germany as well.

Landline telephone

In the homeland of the asylum seekers, the landline telephone was used more frequently than in the target country Germany (Figure 10.26). Furthermore, compared to Germany, more asylum seekers and refugees have used the landline telephone in their home country. Many of the survey respondents answered they never use the landline telephone in Germany (63.64%; N=11), while for the home country, only 9.09% stated they have not used the landline telephone (N=10), if they ever used it. An amount of 45.45% stated they have used the landline telephone daily in their homeland and an additional 27.27% a few times a week in their homeland. Total 18.18% stated they use the landline telephone on a daily basis in Germany and 9.09% each stated to do so a few times a month or about once a month.

Taking a look at the motives, in the home country of the asylum seekers, most were motivated to use the landline telephone for information and self-presentation (75.00% agreed on each). Furthermore, 66.66% each agreed that entertainment and socializing were also a motive for using the landline telephone in the home country. Regarding the situation in Germany, most people had a neutral opinion (50.00%) on all four motives (information, entertainment, socializing, and self-presentation).

10.4 Traditional/offline media

Books

The most used traditional media by asylum seekers are books. A total of 53.6% stated they have read books in their home country or read books in Germany

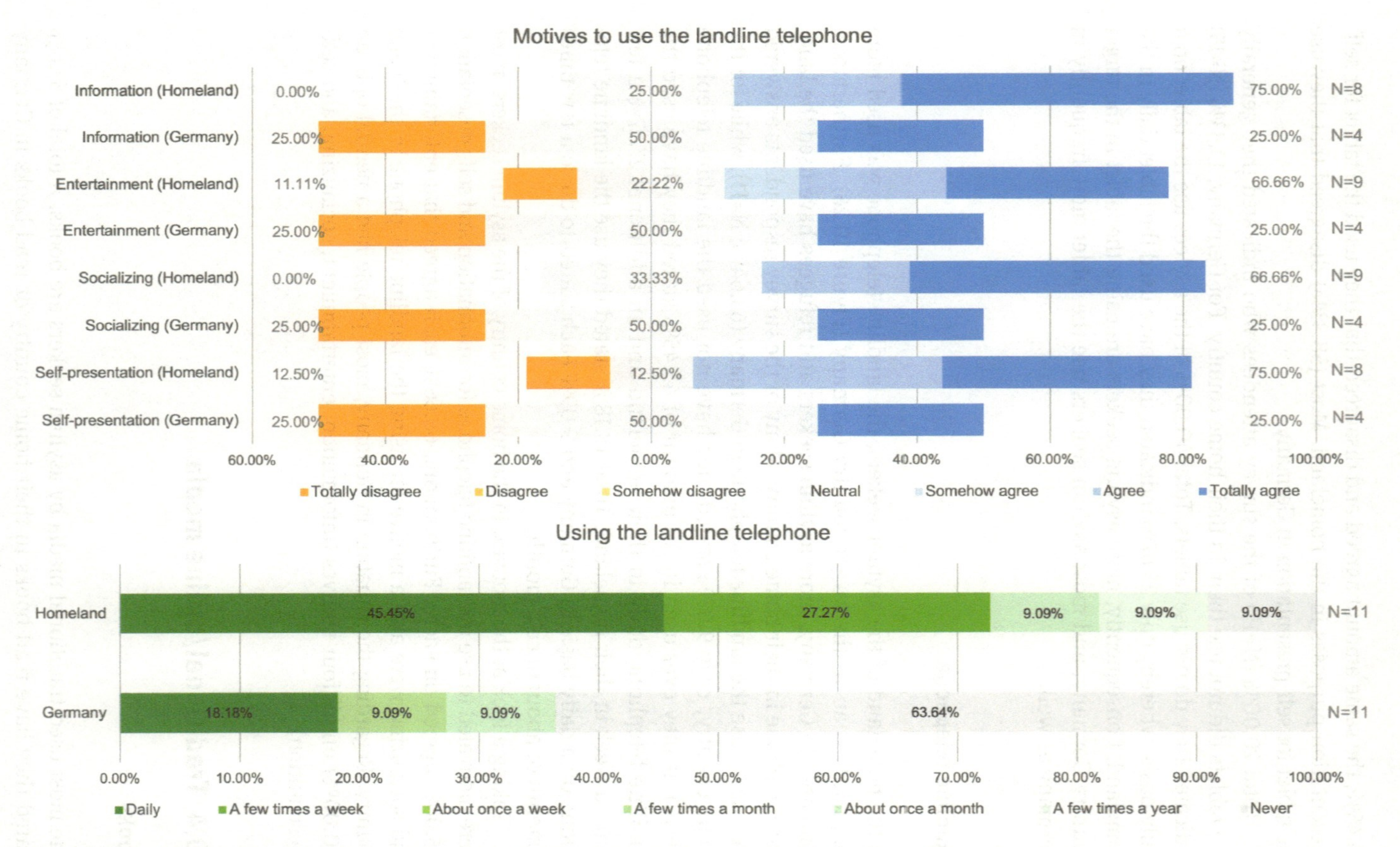

Figure 10.26: Motives to use landline telephones in home country and Germany as well as usage frequency.

(Table 10.2). According to the survey results, asylum seekers have read books more frequently in their homeland than in Germany (Figure 10.27). Most (52.78%; N=36) who said they read books do so on a daily basis in their home country and an additional 16.67% have read books once a week. The frequency of reading books in Germany has decreased a bit compared to the frequency in the home country. Books are read on a daily basis by some (40.00%; N=36) in Germany and an additional 17.14% read books at least a few times a week.

Taking a look at the motives to read books, 68.57% agreed that they have read books in their home country to receive information. For the situation in Germany, still 58.82% agreed to read books for information. Around 55 percent each agreed to read books for socializing in Germany (54.84%) and agreed they have read books for socializing in their home country (55.88%). What motivates asylum seekers slightly more to read books in Germany is to build their personal identity. It is noticeable that the participants expressed a more neutral opinion for all four motives in Germany.

SMS

As shown in Figure 10.28, more asylum seekers have used the opportunity to write SMS in their homeland and they wrote SMS more frequently. Whereby 3.03% (N=33) of the participants who used SMS answered they have not written SMS in their home country, total 11.76% (N=34) stated to not write SMS in Germany. If asylum seekers indicated they have written SMS in their home country, 63.64% did so on a daily basis and an additional 27.27% did so a few times a week, resulting in more than 90.00%. Total 38.24% indicated to write SMS in Germany daily and 17.65% write SMS a few times a week. All in all, SMS is used less frequently in Germany and even fewer asylum seekers write SMS in Germany.

The most important motives for asylum seekers to write SMS are socializing and information, especially while the asylum seekers lived in their home country. Socializing was a motive for 75.86% to write SMS in the home country, whereby 62.96% of the asylum seekers agreed to write SMS for socializing purposes in Germany. Total 74.07% agreed they have written SMS in their homeland to receive information. In Germany, the amount decreased, here, 64.00% agreed to write SMS for information. Around half of the respondents also agree to write SMS for entertainment or for self-presentation. All four motives decreased in Germany.

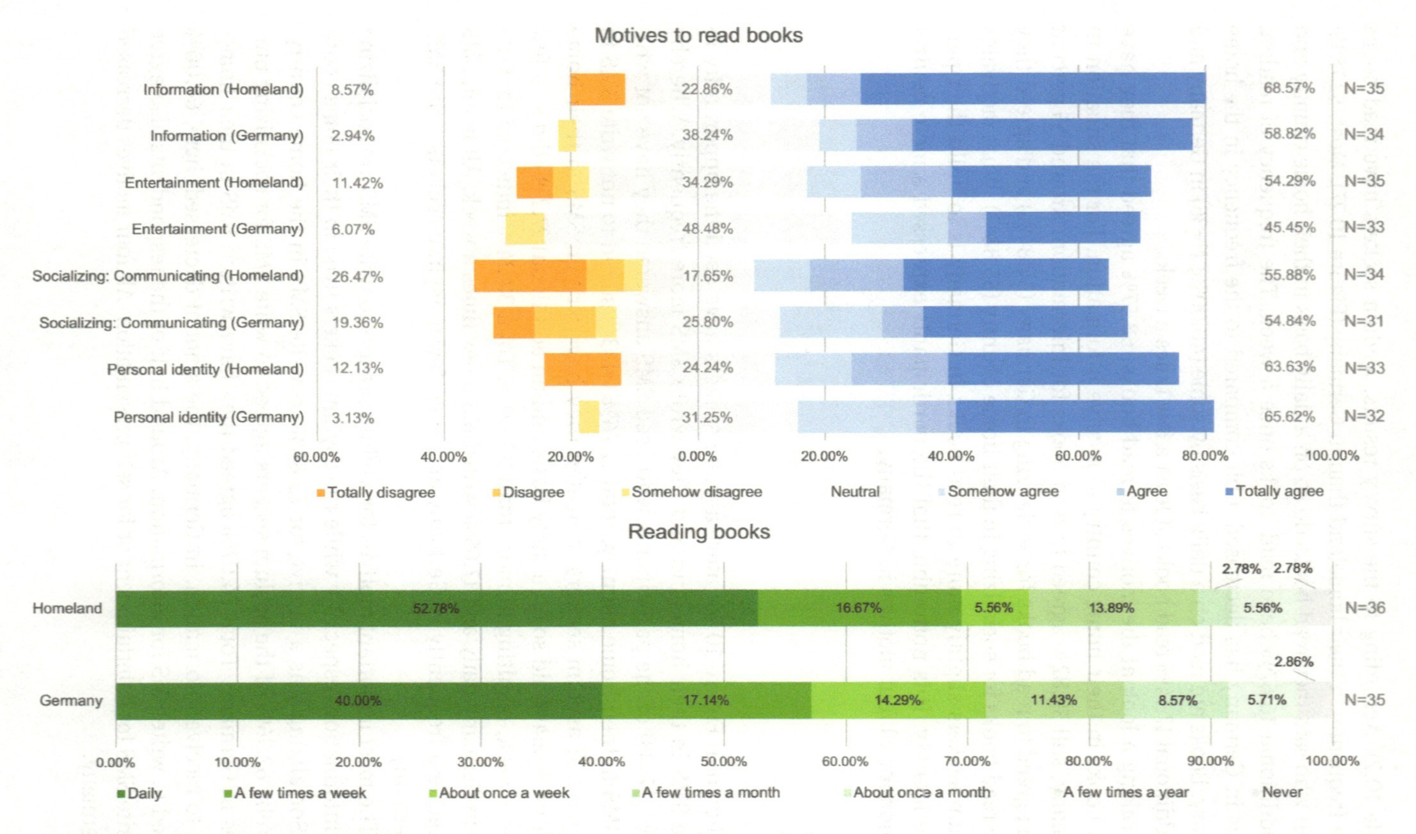

Figure 10.27: Motives to use books in home country and Germany as well as usage frequency.

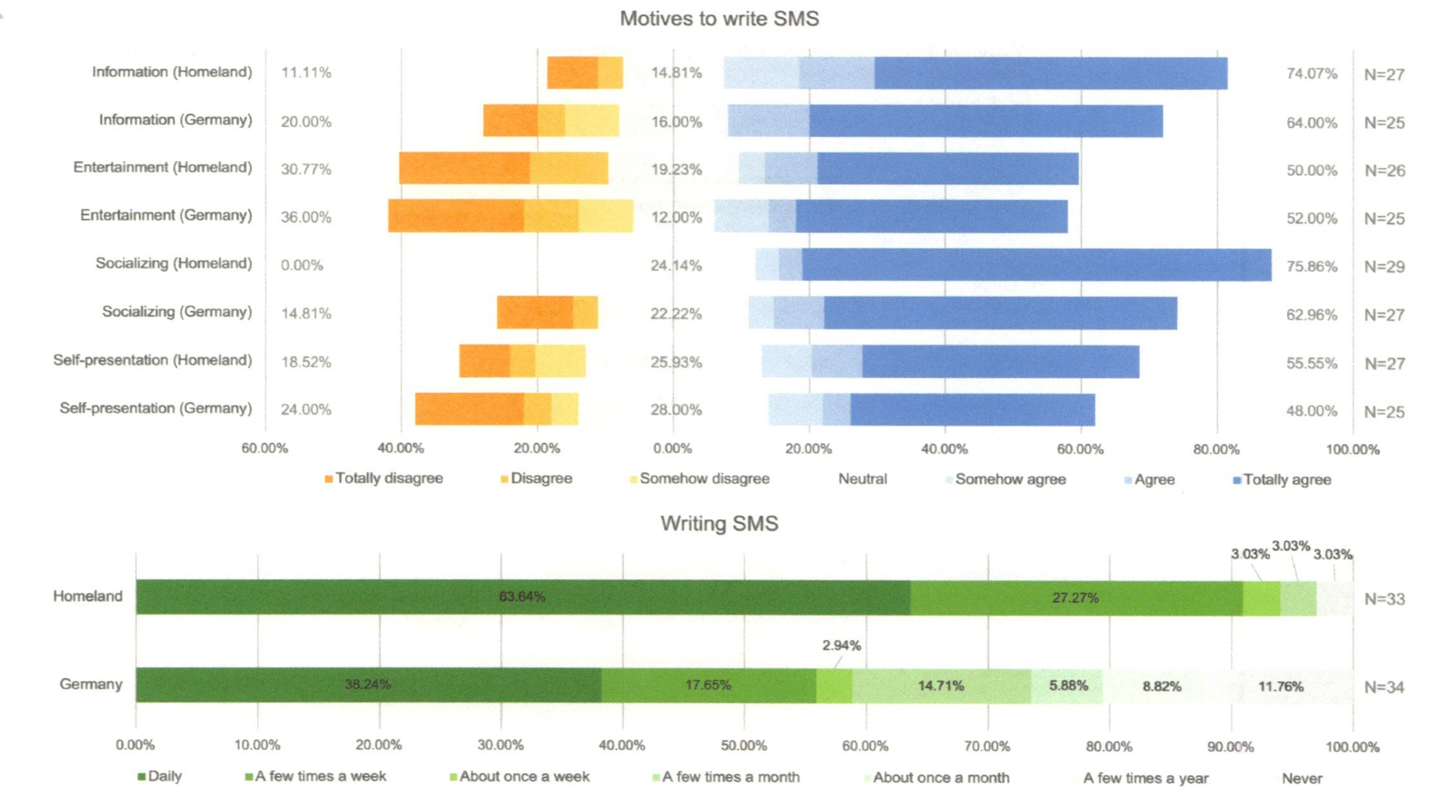

Figure 10.28: Motives to use SMS in home country and Germany as well as usage frequency.

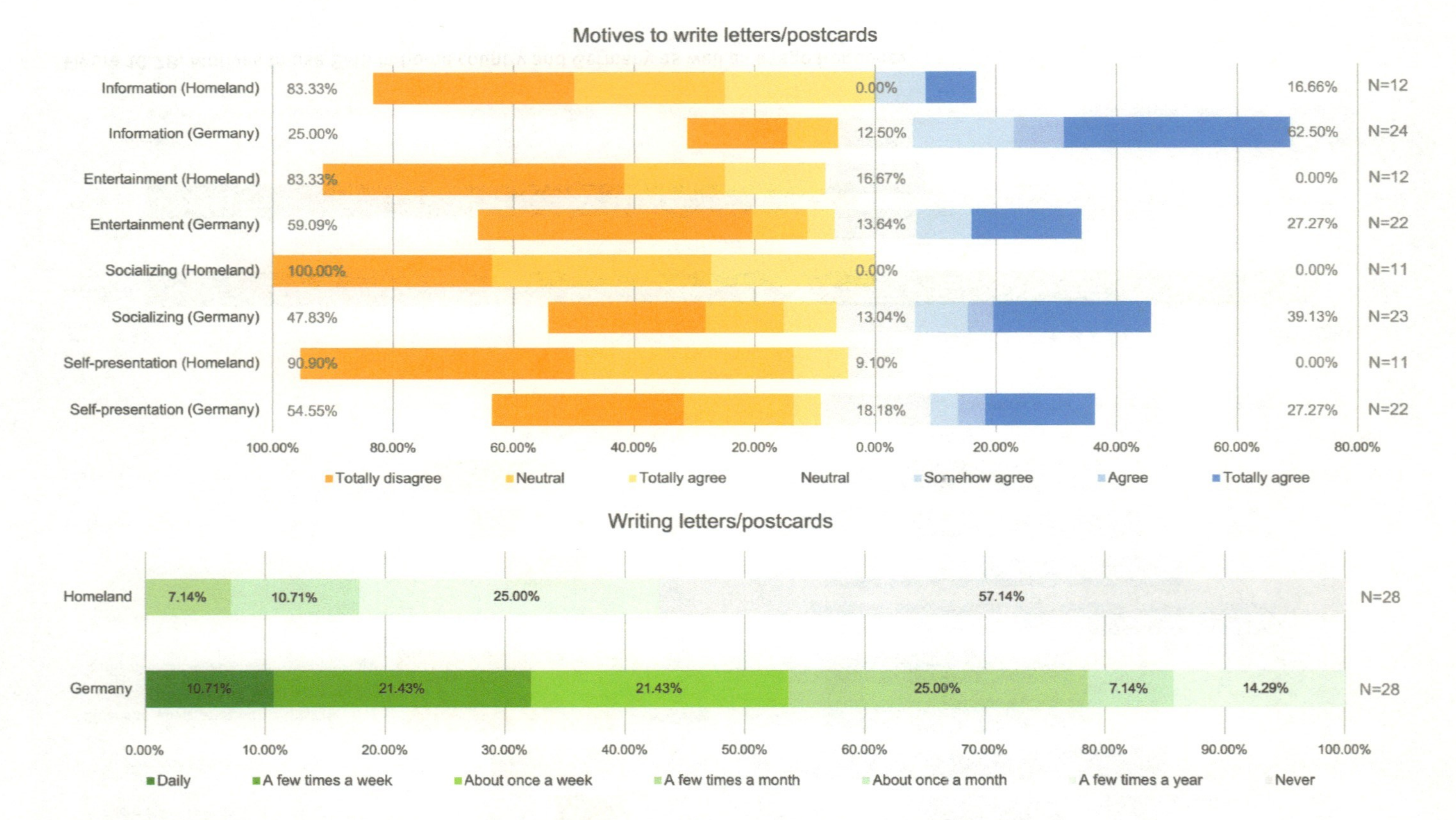

Figure 10.29: Motives to use letters/postcards in home country and Germany as well as usage frequency.

Letters/postcards

According to the results of the survey, a total 57.14% (N=28) indicated they have never written postcards or letters in their home country (Figure 10.29). Those who wrote letters or postcards only did it a few times a month (7.14%) or even less frequently. Total 10.71% stated they did so about once a month and 25.00% stated they have written letters or postcards a few times a year. Nearly 80 percent (N=28) of asylum seekers who write letters or postcards indicated to do so at least a few times a month (25.00%) or more frequently in Germany. More asylum seekers write letters or postcards in Germany and they do it more frequently.

The most important motive to write letters or postcards is information. Only 11.11% agreed and total 83.33% disagreed they have written letters or postcards in their homeland for information, whereas 62.50% of the asylum seekers agreed to write letters or postcards for information in Germany. Although the three motives socializing, entertainment, and self-presentation gained importance as a motive to write letters or postcards in Germany, most asylum seekers and refugees disagree to be motivated by these aspects. Even 100% disagreed that socializing was a motive to write letters or postcards in their home country and 90.00% disagreed on self-presentation as well. All in all, more asylum seekers agreed to be motivated by the four motives since living in Germany.

Newspapers/magazines

Total 90.90% of the asylum seekers who have read newspapers or magazines in their homeland agreed that their motive was information (Figure 10.30). The amount of asylum seekers who agreed that information is their motive to read newspapers or magazines in Germany decreased to 65.00% (around 25 percent points decrease). Information is the main motive for asylum seekers to read newspapers or magazines. To build their personal identity was a motive for nearly 70 percent in the homeland (68.18%), whereas in Germany, only 30.00% agreed and even 35.00% disagreed to read newspapers or magazines in order to form their personal identity. The amount of asylum seekers who read newspapers or magazines for entertainment or socializing also decreased when comparing the amount for the homeland (54.54% entertainment; 63.63% socializing) with the amount for Germany (25.00% entertainment; 35.00% socializing).

Asylum seekers read newspapers or magazines less frequently in Germany. Also, the amount of asylum seekers who read newspapers or magazines in Germany decreased. Total 37.50% (N=24) have read newspapers or magazines on a

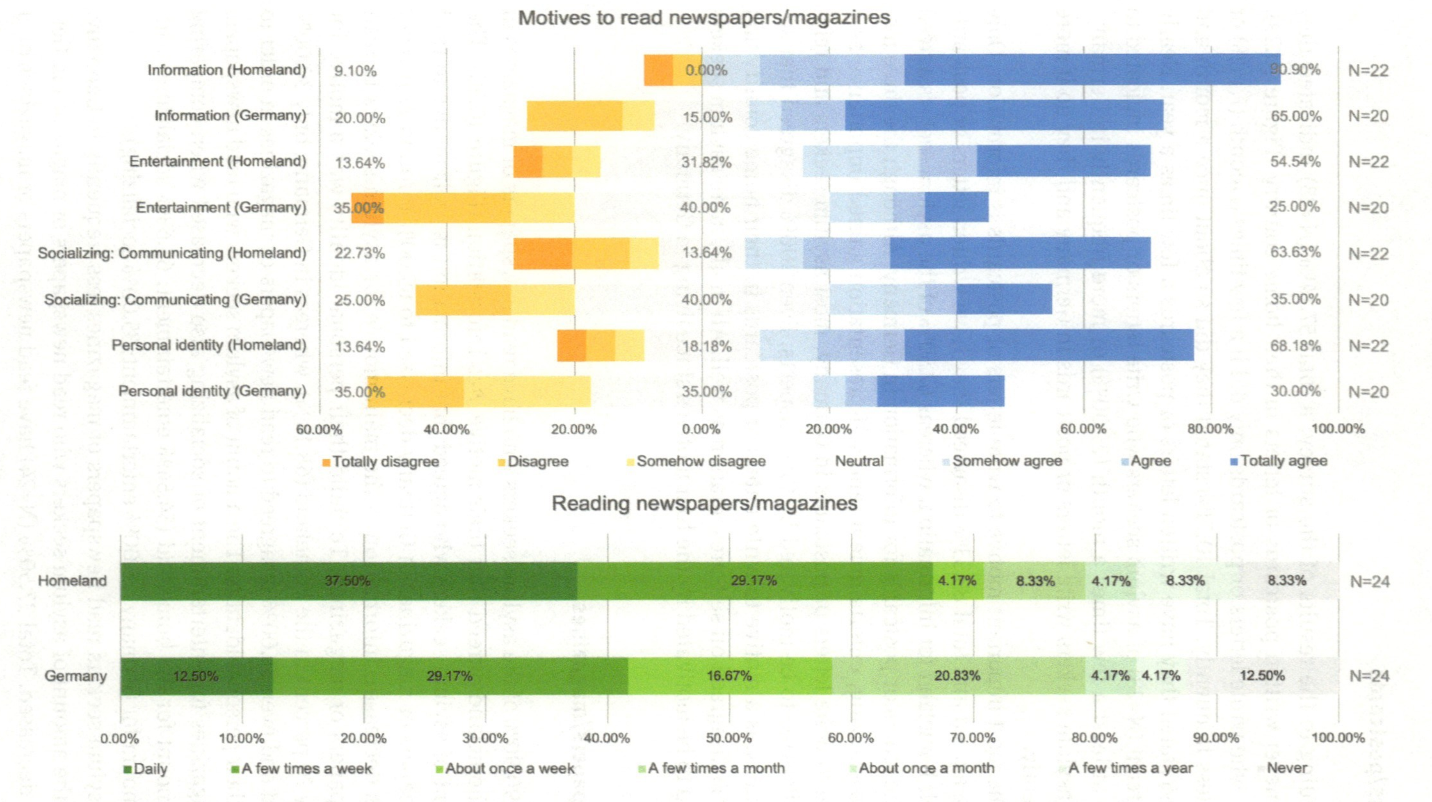

Figure 10.30: Motives to use newspapers/magazines in home country and Germany as well as usage frequency.

daily basis and an additional 29.17% have read newspapers or magazines a few times a week in their homeland. For Germany, only 12.50% (N=24) read newspapers or magazines on a daily basis and also 29.17% read them a few times a week. Total 12.50% stated to not read newspapers or magazines in Germany.

10.5 Conclusion

A survey was conducted in order to collect data on the ICT, social media, and online and traditional media usage of asylum seekers in Germany in comparison to their home country. The majority of survey participants came from Middle Eastern countries, i.e. Syria, Iraq, and from Turkey. Therefore, the mother tongue of most asylum seekers and refugees who attended the survey was Arabic, Farsi, and Turkish. To be able to answer the survey properly, many attendees already had an advanced German language level certificate. It strikes that most respondents are male and on average 32 years old. Most participants also stated they came to Germany in 2015, which also marks the beginning of the refugee crisis.

Among the social media services, the most popular ones are YouTube, Facebook, Instagram as well as messaging services. Many asylum seekers began to use YouTube and Instagram on a daily basis in Germany. YouTube is the most used service and is frequently used for information and entertainment, in the home country and in Germany alike. The same motives apply for Instagram, but additionally, asylees agreed to use it for socializing. Facebook is used by more asylum seekers since living in Germany. While it was especially used for self-presentation in the home country, in Germany, most disagree to be motivated to use Facebook for self-presentation purposes. In general, Facebook is and was used for socializing in Germany and in the homeland. Live streaming services are used for information and entertainment. Compared to the home country, more agreed to be motivated by these motives in Germany. A few more asylees started to use live streaming services in Germany and the ones who use it, do so more frequently. Most participants use messaging services daily. Almost no changes for the usage frequency between Germany and the home country were observed. Messaging services are mostly used by asylum seekers to receive information or to socialize. Some also agreed to use messaging services for entertainment. Pinterest is not widely used, in the homeland and in Germany. Asylum seekers have used and use Pinterest for entertainment, while in Germany it is additionally used for information. The majority stated they have used it about once a week in their home country and those who use it in Germany, do so a few times a week as well. Skype is only used by a few asylees. Making calls via Skype was more frequently done in the home country, mainly for socializing. The service LinkedIn was mostly used for socializing in the home country. It is only used by a few participants and most started to use it since living in Germany. Similar to this, only some asylum seekers use Snapchat and most began to use the platform in Germany. Many agreed they have used Snapchat for entertainment and for socializing purposes which is true for the usage in Germany as well. One-third of those who use Snapchat in Germany, do so on a daily basis. TikTok is only used by a few asylum seekers. Those who used it in their home country agreed to use it for all four motives. But most started to use TikTok in Germany and use it in order to get entertained. Also, the usage frequency of TikTok has increased in Germany. Information is the main motive of asylum seekers why they used Twitter in their homeland and use it in Germany. Only a few participants stated they use dating apps, most started using them in Germany. Only one indicated the usage in the home country, mainly for entertainment. The ones who use dating apps in Germany do so for entertainment, socializing, and self-presentation.

The smartphone is the most used and most important ICT in the life of asylum seekers. In the home country as well as in Germany, nearly all respondents who use it do so daily in Germany and also have used it on a daily basis in the home country. Furthermore, the smartphone is used for all four motives, but above all for information and socializing. The computer, notebook, or laptop are mainly used for information and entertainment, in the homeland and in Germany alike. The usage frequency shows only minor changes. Some asylum seekers use the computer, notebook, or laptop less frequently in Germany and fewer asylum seekers use these devices overall. While the television was used by many asylum seekers on a daily basis in the home country, in Germany, the television is used less frequently. Most agreed to be motivated by information and entertainment in the home country and in Germany, but for Germany, fewer asylees agreed. The landline telephone is used by fewer participants in Germany than in the home country, and if attendees responded to use it in Germany, most use it less frequently. The landline telephone was used to fulfill all four motives in the home country, but especially for information and self-presentation. More asylum seekers started to use the tablet in Germany, they also use it more frequently than in the homeland. While they have used the tablet for self-presentation, entertainment, and information in the home country, they only use it for entertainment and information in Germany.

The most used online media which appeared to be in the top ten of all asked media, devices, or services are e-mail, translation services, and search engines. More asylum seekers started to write e-mails in Germany and they do it more frequently than in the home country as well. While socializing was the motive to write e-mails for most in the homeland, in Germany, most people write e-mails for informative purposes. In general, information and socializing are the motives to write e-mails. Many asylum seekers already used search engines in the homeland and continue to use them in Germany. In Germany, search engines are used more frequently, for many, almost on a daily basis. In the home country, many have used search engines on a daily basis as well. Translation services are a very important kind of online service for most asylum seekers in Germany. These services are used on a daily basis by many, starting the usage in Germany. A few participants already used them on a daily basis in the home country as well. Information and socializing, meaning to be able to talk about topics with other people, are motives to use encyclopedias. They are especially valued for information and knowledge. Entertainment is the main motive to use streaming services, but all gratifications are important here. Many started to use streaming services in Germany. Learning apps are and were used for all four motives, but especially for information in the home country and in Germany. More asylees started to use learning apps in Germany and they use them more often. A few more participants began to use news websites in Germany. In the home country, the main motive to use them was information, followed by socializing in order to communicate with others and to build their personal identity. In Germany, asylees are still motivated by the mentioned aspects, but fewer participants agreed with this. Entertainment is a reason to use news websites as well. More asylum seekers played digital games in their homeland and they also played them more frequently. Most even did so on a daily basis. While most asylees who still play digital games do it on a daily basis in Germany, some do it less frequently and some even quit playing digital games. The main motive to play digital games is entertainment. Even more asylees agreed to play digital games for entertainment in Germany than in the home country. In the asylum seekers' homeland, podcasts were used mainly for entertainment and for information. In Germany, the main motives are information followed by personal identity, but asylees are also motivated to listen to podcasts in Germany in order to socialize, to talk with other people about the content. Some asylum seekers started listening to podcasts in Germany and most do so a few times a week. In the home country, those who listen to podcasts did it a few times a week too.

For traditional/offline media, SMS are written by fewer asylum seekers and refugees in Germany than in their home country, and if respondents use it in Germany, most stated to write SMS less frequently. SMS is used for all four motives, information, entertainment, socializing, and self-presentation but above all for information and socializing. Books and newspapers or magazines are read by nearly the same amount in Germany as in the homeland, but less frequently. Asylum seekers are motivated by all for motives, but particularly to get informed and to build their personal identity. Books are the most used traditional media followed by SMS. However, especially letters are written or read by more asylum seekers in Germany compared to the home country and even more frequently. The motives of the respondents to write or read letters in Germany are to receive information and to socialize. Asylum seekers who used traditional media in their home country mostly still use them in Germany, but less frequently. An exception is writing letters. A possible explanation would be that more asylum seekers write letters to communicate with German authorities.

When asylum seekers were asked what they think are the reasons why their behavior in Germany changed in comparison to their behavior in their home country, many answered that learning the language of the new home country or caring for their family are important factors.

11 Information needs of asylum seekers in a new country

What are the questions asked by asylum seekers in online forums? What kind of topics are of interest to them? What do they need information on and what kind of situations do they need assistance with? To answer this, questions by refugees and asylum seekers asked in an online forum called *Wefugees* were analyzed. To this end, a content analysis was conducted. The intention of the online platform Wefugees is the support of asylum seekers with their information needs concerning various topics, especially about living in Germany (Figure 11.1). Users have the opportunity to ask questions and find answers through previous asked questions of other users. Experienced volunteers and experts provide answers to the questions and help to find solutions. An interactive community for everyone, refugees and volunteers alike, is the goal. The shared knowledge is collected and stored, and therefore accessible and available for everyone.

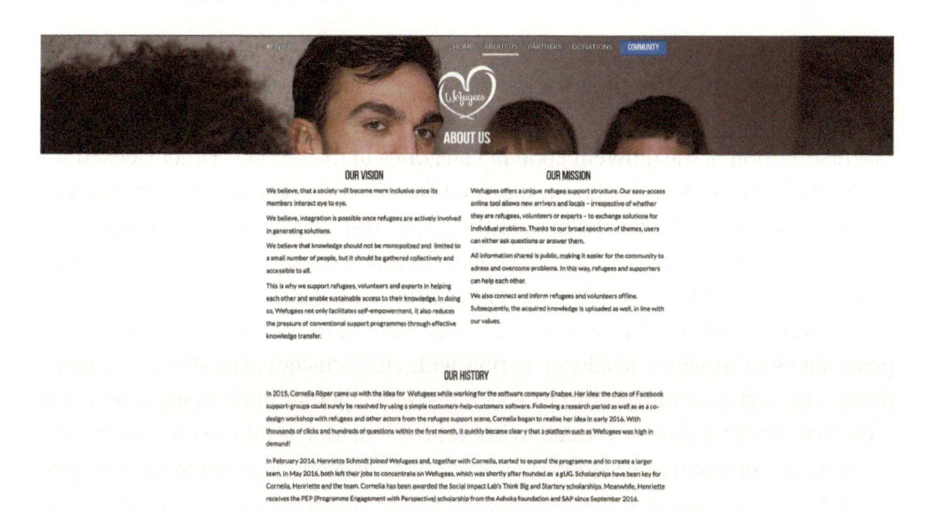

Figure 11.1: About us page Wefugees forum.

This content analysis considered 139 questions, not including questions asked by volunteers. The questions were analyzed intellectually by the two researchers. The content analysis was conducted on the category "Other Questions" from the Wefugees forum, since the website provides predefined categories for sorting and assigning questions, namely: legal advice, asylum proceedings, education, home

https://doi.org/10.1515/9783110672022-011

and living, work, activities, information and offers, money, healthcare, and "How can I help." Overall, a number of 23 content categories (Table 11.2) with additional 17 document/legal status categories (Table 11.1) were formed during the coding process. For some questions, multiple assignments were made. Giving an example, if someone asked about marriage in context with traveling, the question was coded with both categories, "marriage" and "traveling." Therefore, sometimes, a few of the examples provided are used for different categories. Some questions asked by asylum seekers are displayed. Since most of the questions contained some grammatical errors in English or German, some of the quotations included for demonstration have been modified. To be able to display the information needs of asylum seekers in an appropriate way, the numbers of questions asked in the pre-formulated categories are presented as well (Table 11.3). Here, two states of data display the development of the information needs of refugees over time. Finally, some further details about the requests in these categories are added to this chapter. Based on this, the following research question (RQ) will be explored:

RQ4: What kind of information do asylum seekers search for in online forums?

11.1 Documents/legal status

The distribution of the different content categories in the section "Other Questions" of the Wefugees forum displays a big need for support regarding documents and the asylum applicants' legal status (Table 11.2). The largest number of questions and requests was identified here (43.88%). Taking a look at Table 11.1, the document type that is the most often inquired about is the residence permit (36.07%). For example, a father of a German daughter wanted to inform himself about the possibilities to receive a residence permit with child custody. Questions regarding living with a child and trying to receive a residence permit in Germany were often identified. Another person wanted to know how long one is allowed to stay in Germany with a permanent residence permit card. A different question in the category residence permit was asked about traveling. "Can I travel to other E.U. countries with my 6 months' residence permit sticker?" someone asked.

Following, there were also comparatively a lot of requests about the driver's license (18.03%). The requests in this category concerned how to apply for a driver's license or how to receive driving lessons in English. Also, someone wanted to know how to get a driver's license for a motorcycle. Following, further questions identified were about the visa (9.84%). For the category visa, many questions were related to marriage (e.g., "Should we register our marriage in Germany before applying for a family-unification visa?") or related to

Table 11.1: Number of questions assigned per documents/legal status category.

Category	Number of questions (N=61)	
	Absolute frequency	Relative frequency
Residence permit	22	36.07%
Driver's license	11	18.03%
Visa	6	9.84%
Asylum application	4	6.56%
Temporary suspension of deportation	4	6.56%
Ban on deportation	3	4.92%
Entry permit	2	3.28%
Birth certificate	2	3.28%
ID	2	3.28%
Deportation	2	3.28%
Refugee status	2	3.28%
Marriage certificate	2	3.28%
Registration certificate	2	3.28%
Police clearance certificate	1	1.64%
Subsidiary protection	1	1.64%
Letter of recognition	1	1.64%
Rejected asylum application	1	1.64%

traveling (e.g., "I have a blue pass (three years), do I need a Ukrainian visa to travel to Ukraine?"). Other important identified topics are the asylum application (6.56%) and the temporary suspension of deportation (*Duldung*) (6.56%). One request concerned the question if asylum seekers with a temporary suspension of deportation can open a bank account in Germany. Someone else wanted to know if it was possible to marry a German woman in India while having the toleration status in Germany. The ban on deportation (4.92%) also concerns questions about documents or the legal status. To give an example, a person received a ban on deportation and firstly, wanted to know who to contact at a job center. Furthermore, the person wanted to know if it was possible to start a training position for work and to live in a private accommodation. Other questions were about the entry permit, birth certificates, the ID, deportation, refugee status, marriage certificates, or registration certificate (3.28% each). For the entry permit, one person needed advice on how to get an entry permit as an asylum seeker from Yemen. Another question asked how long the process takes to verify a birth certificate in Germany. Further topics asked about are police clearance certificates, subsidiary protection, letter of recognition, or a rejected asylum application (1.64% each).

11.2 Employment/job

The requests from the asylum seekers and refugees on Wefugees often deal with different topics in relation to employment and jobs (13.67%). One user of the forum asked if it is allowed and possible for refugees to volunteer in Germany. Another user wanted to have advice if a master's degree should be completed or work started first: "Should I complete my master of science in Germany or search for employment as an experienced electrical engineer?" the person asked. Furthermore, employment is often inquired about in relation to the legal or asylum status (e.g., "Can rejected asylum seekers work in Germany?" "I have the Duldung status – Am I allowed to have a trading license?" "Is it legal that the Foreigners Registration Office didn't issue a permission to work, despite a deportation ban?" and "Does accepting a job change my legal status?") "I have applied for a residence permit. When will I be allowed to work?" someone wanted to get information about. Thus, the refugees and asylum seekers have an information need for clarification on work permissions and regulations. One person asked for advice regarding his or her employer – the minimum wage was not paid and taking leave was not permitted. However, another user needed information about rights as an employer in Germany. Many of the requests also deal with work and training certificates or documents. "How can I work as a carpenter without any certificates or training?" one asylum seeker wanted to know. Another request was related to lost papers and how to certify them: "I lost the papers that prove my training, how can I certify it?" (see Figure 11.2). Further inquired about was, if a permission for employment can be applied for and if any papers are necessary for the procedure. Another user asked if the social welfare office or the federal office for migration and refugees will still pay alimony if an unsalaried training position is taken. In relation to this, it was also of interest what financial benefits are received from the welfare office if one has a mini job. Finally, a user asked for support from German authorities during the job researching process.

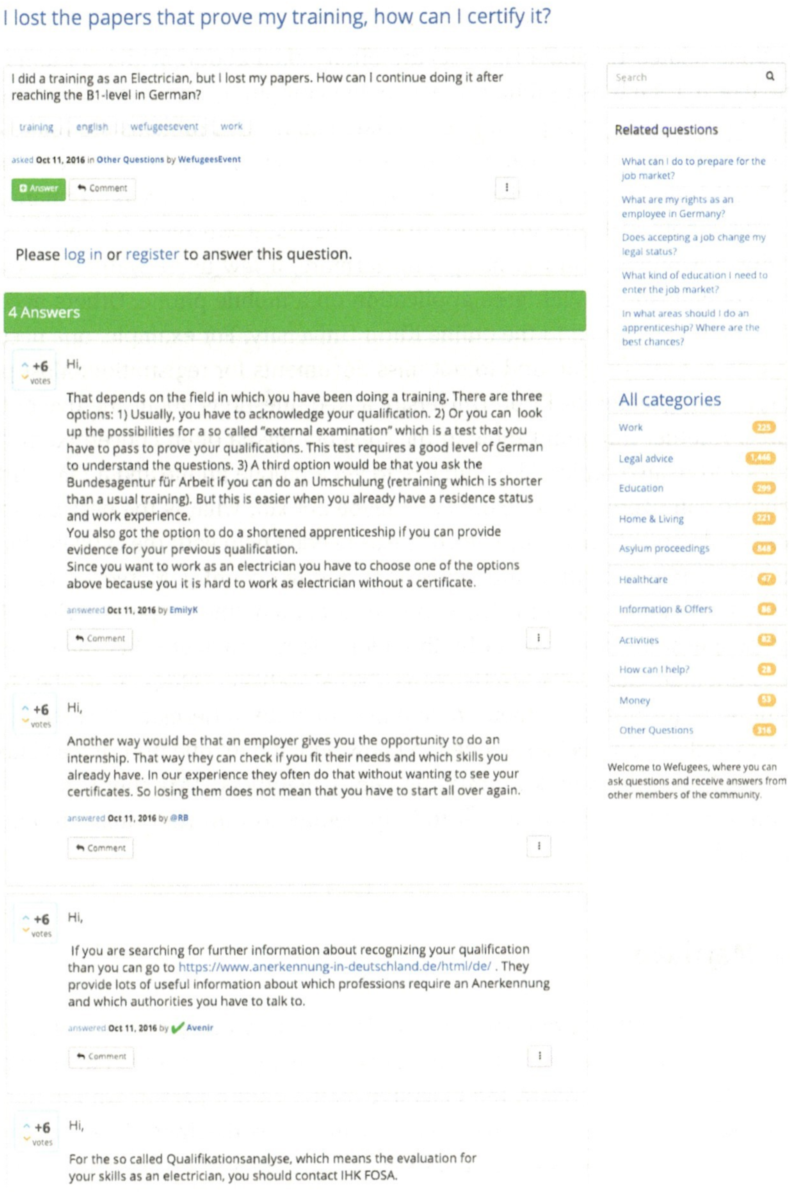

I lost the papers that prove my training, how can I certify it?

I did a training as an Electrician, but I lost my papers. How can I continue doing it after reaching the B1-level in German?

training | english | wefugeesevent | work

asked Oct 11, 2016 in Other Questions by WefugeesEvent

Answer | Comment

Please log in or register to answer this question.

4 Answers

+6 votes

Hi,

That depends on the field in which you have been doing a training. There are three options: 1) Usually, you have to acknowledge your qualification. 2) Or you can look up the possibilities for a so called "external examination" which is a test that you have to pass to prove your qualifications. This test requires a good level of German to understand the questions. 3) A third option would be that you ask the Bundesagentur für Arbeit if you can do an Umschulung (retraining which is shorter than a usual training). But this is easier when you already have a residence status and work experience.
You also got the option to do a shortened apprenticeship if you can provide evidence for your previous qualification.
Since you want to work as an electrician you have to choose one of the options above because you it is hard to work as electrician without a certificate.

answered Oct 11, 2016 by EmilyK

Comment

+6 votes

Hi,

Another way would be that an employer gives you the opportunity to do an internship. That way they can check if you fit their needs and which skills you already have. In our experience they often do that without wanting to see your certificates. So losing them does not mean that you have to start all over again.

answered Oct 11, 2016 by @RB

Comment

+6 votes

Hi,

If you are searching for further information about recognizing your qualification than you can go to https://www.anerkennung-in-deutschland.de/html/de/ . They provide lots of useful information about which professions require an Anerkennung and which authorities you have to talk to.

answered Oct 11, 2016 by ✔ Avenir

Comment

+6 votes

Hi,

For the so called Qualifikationsanalyse, which means the evaluation for your skills as an electrician, you should contact IHK FOSA.

Search

Related questions

What can I do to prepare for the job market?

What are my rights as an employee in Germany?

Does accepting a job change my legal status?

What kind of education I need to enter the job market?

In what areas should I do an apprenticeship? Where are the best chances?

All categories

Category	Count
Work	225
Legal advice	1,446
Education	299
Home & Living	221
Asylum proceedings	848
Healthcare	47
Information & Offers	86
Activities	82
How can I help?	28
Money	53
Other Questions	318

Welcome to Wefugees, where you can ask questions and receive answers from other members of the community.

Figure 11.2: Sample question about lost training certificates and related answers.

11.3 Media

Questions related to media (12.95%) are often asked about. Media contains aspects like ICT such as mobile phones or the computer, online media such as YouTube, websites, or apps and other related subjects. Problems are related to German websites and translations of them, as someone asked: "Where can I find English or Arabic translations or translators? I do not speak German and all refugee help websites are in German." The issue was even related to help websites for refugees in Germany. Furthermore, a few users wanted to know how to download the Wefugees application on a mobile phone. Others needed advice on issues related to the online Kiron University. For example, one needed help with the registration and to not miss documents for registration. Most people already registered at Kiron University asked questions about the live online courses. Another important topic in the category media often concerned mobile phones and related topics. "I would like to get a mobile phone and a SIM card to be able to call people in Germany and maybe abroad, what is the best/cheapest place I could go to?" Whereby someone else needed help with the cancellation of their mobile phone contract after moving to another country. Moreover, one person wanted to know about the trustworthiness of an online support service. Also, some people were interested in finding blogs about cities or other media to be able to orientate themselves in Germany. "What websites, blogs, YouTube channels or other mediums are there to help me orientate in Germany?" some of the refugees and asylum seekers wanted to know. Finally, for one user, it was important to help people to look for missing family members: "Are there web pages or numbers I can look up or call to help people to find their missing family members?"

11.4 Marriage

Marriage is also often inquired about (12.23%). Nearly every question from this category was also assigned to the category document/legal status. Furthermore, the questions often concern the marriage status with a person who is not an asylum seeker or refugee, or whose application was declined. In some of the analyzed requests, one person lives in another country than the partner or the partner still lives in the home country. Sometimes, both people have applied for asylum in two separate countries and want to get married. A refugee who received a residence permit in Italy wanted to get married to a German woman living in Germany. They wanted to process their marriage in Italy and needed help with the single status certification in order to be able to marry in Italy.

Someone else who got married to a German woman wanted to know why he still needs to file for asylum. Another asylum seeker in Germany wanted to marry her boyfriend who lives in the Middle East and asked for information on how to proceed. A refugee with a temporary suspension of deportation wants to marry a German. Because the person did not have a passport or ID, the question was, if it was possible to get married in India. Furthermore, a Latvian woman wanted to inform herself about the documents she will need in order to be able to marry a refugee who lives in and is registered in Germany. Another user of the Wefugees forum who is an Iraqi refugee in Germany wanted to marry a person from the U.K. The question was about if the person would be allowed to travel to the U.K. after the marriage to visit the family regularly. One refugee in the U.K. wanted to marry a refugee who is living in Germany. The person wanted to know if the refugee status granted in Germany is still applicable after moving to the United Kingdom. "I am a refugee in Munich, Germany. I have a three years' residence permit in Germany. I want to marry my girlfriend, but she is a refugee without documents and she lives in a French refugee asylum home. How can I marry my girlfriend so we both can live together in Germany?" another refugee asked. It seems to be an unusual circumstance, but for some refugees, it is an important information desire in their life. Someone's fiancé is living in Nigeria and they wanted to get married in Germany and were waiting for four months for a visa. The question was, how long will it take to get a visa in order to get married. "Should we register our marriage in Germany before applying for a family-unification visa?" was another question. For some people it is important to know how to apply for a residence permit after marriage. Another person was interested in the procedure: "What is the procedure and requirement for a permanent residence permit through marriage?" Some requests were related to the marriage certificate. Someone's marriage certificate was translated and therefore the person wanted to know how to verify it. In another case, the foreigners federal office took the unmarried certificate and the marriage status certificate was kept by the marriage office.

11.5 Travel

If the travel aspect (7.19%) is concerned, all inquiries are related to the question if it is allowed to visit other countries in Europe when some form of protection or a legal or asylum status in Germany has been granted. "Which countries can I visit without a visa with my three years' limit residence permit and travel ID (Geneva convention)? EU countries? Schengen countries? And, can I go to Switzerland or Norway?" was one question. Another user wanted to know which countries are allowed to visit if one also holds a passport and has a three years' residence permit. The same question was asked for a six-month permission. Another user wanted to know if a visa to travel to the Ukraine is needed. And even someone wanted to know if it is allowed to travel to the home country while having a permission to reside in Germany. A further question was asked by a refugee who is a university student in Germany and exchanged to Sweden. The student has to apply for a residence permit there and wants to know if one is allowed to travel to Iceland. It was furthermore requested by someone who is from Syria and has received a residence permit in Germany to get help to return to Turkey. Some requests were related to marriage. An Iraqi refugee wanted to marry a German and asked if it is possible to travel to the United Kingdom. Someone else has a residence permit in Italy, got married in Denmark, and wanted to travel to the U.K.

11.6 Language

Questions related to language are a common subject for asylum seekers and refugees (6.47%). Most users have difficulties understanding the language, especially when filling in administrative documents or reading letters from the authorities and require help with translations. Therefore, someone was in search of a cheap translation service in Berlin. One user had questions about a German text regarding the residence permit application. Another one needed help with the translation of English sentences into German, but did not include further details. Someone added a request about finding a translation service in Munich to translate texts from Arabic to English. Another asylum seeker claimed that all websites designed to be helpful for refugees are in German and that English or Arabic translations are needed to be able to understand the information displayed. "I am searching for a language café" a user wrote as a request. Language cafés are coffee shops offering meetings for people who want to learn a language through conversations with others, sometimes with native speakers.

11.7 Institutions

Some information is inquired regarding the subject institutions (6.47%). Most analyzed requests were related to the foreigners' registration office. Someone claimed that the foreigners' registration office did not answer requests and if they do, the response is late. Therefore, the person wanted to know how to speed up the process of their asylum application. Another question was in relation to a visa, as the foreigners' registration office told the person to get a visa from the home country to visit their children. “How can it be that I have to go back to my home country in order to get a visa?” was the exact question. “Hey I need to go to the Syrian Embassy, what is the address and how can I go there?” was asked by another user. One refugee or asylum seeker was wondering about the conditions of German refugee centers: “What are the conditions of the refugee centers in Germany?” A few questions also involved employment. A refugee needed support from German authorities during research for a job position. Someone else, who also asked about jobs, wanted to know who the responsible contact person at a job center is if ones lives in an accommodation, holds a training position and a ban on deportation was granted. Last but not least, for the category institutions, someone needed information about organizations and programs to get back to their home country.

11.8 Studying

When analyzing the questions from the category studying (6.47%), nearly every question dealt with Kiron University. It is an online university that was funded to make access to higher education easier for refugees. Many questions were asked about how to successfully register at Kiron University or on how to find or even attend courses. One person who asked about courses wanted to know details about the social sciences classes at Kiron University. “Who is taking courses in social sciences at Kiron University, I want to ask some questions” one requested. Someone else was not able to find the English courses: “I am a student at the Kiron University but I cannot find the English course, any help?” Another request was regarding a German online live course. Basic information updates were missing regarding this course. Someone else had problems with a finished course at Kiron University as the system did not display the status of attendance. For those who wanted to apply to Kiron University, one person with a bachelor’s degree in economics received in Damascus wanted to apply for a master’s degree at Kiron University and needed help with the application. In another case, a refugee from Iraq who stayed in a camp and did not have a residence permit in Germany wanted to know what documents one has to submit to Kiron University. Further problems were observed regarding the available university seats in Germany and in partner universities: “I’ve read the Q&A of the Kiron University and I’ve seen that the number of seats at the partner universities is limited. Is it possible that after two years of studying online, I won’t get into one of them? What are the options I would have left in that case?” The only question not related to Kiron University was asked by someone who needed advice if one should apply for a master’s degree in Germany or rather search for a job in order to gain experience.

11.9 Paternity rights

Paternity rights have been mentioned sometimes and are a topic of interest as well (5.76%). Questions were often related to pregnancy or travelling to a certain country in order to visit children. A woman in Germany wanted to know if it is possible to be deported when one is pregnant from a German man. Another one, whose wife is pregnant and lives in Germany, needed advice on how to be able to be by her side when she gives birth as he is living in another country. Someone whose children live in Germany has applied for refugee status in Greece and wanted to know how to get to Germany and if one is allowed to stay in Germany because of the children. Another refugee married a Polish wife in Germany with whom he has a child and wanted to know if it is possible to receive a residence permit in Germany. Someone was not sure about the rules in Germany when a child is born concerning surnames. "Which last name does a newborn child get?" the person asked. Another one who received child custody wanted to get information on changing the surname of the child. Others asked questions about issues regarding the residence permit while having child custody for a German child or needed information about how long the verification of a birth certificate at the registry office will take.

11.10 Money

Social benefits and payments from the social welfare department are often a topic involved in the category money (5.04%). One user sought information about the amount of money the job center will pay for an accommodation in Düsseldorf. Someone else who also needed advice on accommodation and money wanted to know if the job center pays the apartment security deposit. Another one wanted to know the amount of social benefits one will receive or what other help will be provided from the social welfare department or federal office for migration and refugees when working in an unpaid training position. Similar to that, a question was asked in relation to a mini job. "What help will I still get or not get from the federal welfare office when I start a mini job?" was asked by that person. One user had an issue with their job salary. The employer did not want to pay the minimum wage. Therefore, the user wanted to know where to get help. Someone inquired if they would be able to open a bank account if they received a temporary suspension of deportation status. Someone else was interested in crowdfunding, they wanted to get feedback on experiences with crowdfunding from other users at the forum – asylum seekers and volunteers alike.

Table 11.2: Content categories of Wefugees forum. Sometimes multiple assignments.

Category	Number of questions (N=139)	
	Absolute frequency	**Relative frequency**
Documents/legal status	61	43.88%
Employment/job	19	13.67%
Media	18	12.95%
Marriage	17	12.23%
Travel	10	7.19%
Language	9	6.47%
Institutions	9	6.47%
Studying	9	6.47%
Paternity rights	8	5.76%
Money	7	5.04%
Accommodation	5	3.60%
Integration into German culture	5	3.60%
Supplies	4	2.88%
Travel routes	3	2.16%
Leisure time	3	2.16%
Social contacts	3	2.16%
Insurance	2	1.44%
Dating	2	1.44%
Rights	2	1.44%
Missing family	2	1.44%
Family unification	1	0.72%
Interview	1	0.72%
Culture (other)	1	0.72%

11.11 Accommodation

Another multifaceted issue that refugees and asylum seekers need help with are different information needs about accommodation (3.60%). As refugees are sometimes not allowed to work and earn money for accommodation, they receive housing and other social benefits. "How much does the job center pay for a flat in Düsseldorf?" one person asked. A user wanted to know if it is allowed to live in an apartment, without having to live with other people, when a 6-month residence permit was granted. For another user, it was not clear whether one has to re-register at the municipality if they moved from one city to another. Furthermore, someone is registered as a refugee in the U.K. and got married to another refugee in Germany – will the German refugee status still apply after moving to the U.K. To go on, someone who lived in Germany for two years wanted to move back to Egypt and was searching for an institution or an auxiliary program that would provide help. Therefore, moving was a subtopic of accommodation.

11.12 Integration into German culture

A few questions could be identified about integration into German culture (3.60%) and related topics. For example, the questions deal with the meaning of holidays. It was asked: "What is the reformation day?" One person needed information about emergency telephone numbers in Germany. Someone else was curious about what German citizens think about people from Palestine. Also, the idiomatic expression to press one's thumbs (*Daumen drücken*) in German, similar to *fingers crossed* in English, was one question. What qualifications are needed to attend an integration course was also related to learning about German culture.

11.13 Other categories

Other categories with only a few requests are discussed in this paragraph. In the category supplies (2.88%) the questions concerned where to buy or get supplies. One asylum seeker wanted to know where to get a stroller for a toddler. Where to buy a car when in the middle of the asylum process, was a similar question. Someone else wanted to buy a phone. Another user of the forum asked for an organization that would help to buy an affordable bicycle. Escape routes, leisure time, and social contacts are asked about as well (2.16% each). In the category escape routes, one question was about how to escape from Syria. "What can I do if I am fleeing with my pet?" was asked by someone. In contrast to the requests from people who need advice about escape routes, some people have questions about their leisure time. One person wanted to get information about nearby mosques and where to find them. Someone who seems to be interested in soccer asked: “Is it legal to watch the Champions League or the Spanish League through the Internet?” And a woman wanted to get information about options for women’s-only swimming opportunities. Furthermore, insurance, dating, rights, and missing family (1.44% each) and finally, the topics family unification, interview, and culture (other) were also present (0.72% each).

Table 11.3: Pre-formulated content categories (1 = September 20, 2019; 2 = April 19, 2021).

Rank	Topic	Number of questions 2019 (1)	Number of questions 2021 (2)
1.	Legal advice	891	1,434 (+543)
2.	Asylum proceedings	552	843 (+291)
3.	Education	216	299 (+83)
4.	Home & living	196	221 (+25)
5.	Work	160	218 (+58)
6.	Activities	76	82 (+6; 7th rank)
7.	Information & offers	63	85 (+22; 6th rank)
8.	Money	47	52 (+5)
9.	Healthcare	35	45 (+10)
10.	How can I help?	24	28 (+4)

11.14 Pre-formulated categories

For the existing categories on the Wefugees website (Table 11.3), two time periods of the collected data are presented. The data of the first column (number of questions 1) was collected on September 20, 2019 and the second one (number of questions 2) on April 19, 2021. As the increase of questions in various categories shows, there still remain many questions and information needs of asylum seekers. The category with the most asked questions is legal advice with a number of 1,434 questions. For this category, an increase of 543 questions was observed. In second place is the category asylum proceedings. A number of 843 questions were asked in this category. Here, the number has increased by 291 questions. For the category education, 299 questions are categorized. Education and work (218 questions) both gained a comparatively high number of questions in the documented time span. The category information & offers (then: 63 questions; now: 85 questions) overtook the category activities (then: 76 questions; now: 82 questions). The least important categories and topics are money (52 questions), healthcare (45 questions), and "How can I help?" (28 questions).

12 Experts' insights – Difficulties and information practices of asylum seekers

To reevaluate the findings of this multi methodological study and to get an extended and broader understanding about asylum seekers' application of different ICT and media, including social media, as well as their information behavior, four experts were interviewed to report about their experiences and impressions on asylum seekers and their information behavior, information needs, and information literacy as well as their ICT and (social) media usage.

Each interview partner is an expert in a different thematic area. Starting with Prof. Dr. Carola Richter (IE 1), she and her team evaluated and studied the ICT and media usage and information behavior of asylum seekers before, during, and after their flight. Furthermore, she studied the development of the asylum crisis in Germany. Dr. Juliane Stiller and Dr. Violeta Trkulja (IE 2/3) studied the information seeking behavior and digital literacy skills of refugees during job orientation. They founded their institution *You, We & Digital* in Berlin, Germany which offers workshops to teach digital literacy and information literacy related skills, including workshops especially for asylum seekers. The statements made by Prof. Dr. Carola Richter, Dr. Juliane Stiller, and Dr. Violeta Trkulja are therefore mainly from the German perspective. Prof. Dr. Rianne Dekker (IE 4) studied the role of ICT and communication channels in migrant networks. Further, she and her team investigated the role of social media as an information source for deciding whether and where to migrate. Her investigations and therefore her reports are mainly about the situation in and experiences from the Netherlands. Based on this, the following research question will be explored:

RQ5: What are experts' insights on the information behavior as well as ICT and media usage of asylum seekers?

The experts reported on various difficulties and problems asylum seekers, refugees as well as the federal offices had to face during the refugee crisis (chapter 12.1 Difficulties during forced migration and integration). The asylum seekers' information behavior, including their information needs, used information channels and sources, and information literacy (chapter 12.2 Information behavior) and further, topics related to ICT and general as well as social media are discussed (chapter 12.3 ICT and media; chapter 12.4 Social media). Finally, gender-dependent and age-related differences are explained (chapter 12.5 Age- and gender-dependent differences), followed by a brief conclusion. Some of the results may be repeated as they fit the thematic focus of the various paragraphs.

https://doi.org/10.1515/9783110672022-012

12.1 Difficulties during forced migration and integration

When deciding to flee their home country to seek asylum in another country, asylum seekers use standard channels to receive information, speaking of radios, television, and the Internet. Especially if decided to seek asylum in Germany, the German international broadcasting service *Deutsche Welle* was a first point of contact and a primary source for information (IE 1). During the flight, smartphones were used to navigate and to plan further escape routes. Nearly every person who fled their former home country had a mobile phone. Whether people owned a smartphone or a cellphone depended on their financial means. Those with the greatest digital capital became opinion leaders during escape. This was usually a male person – this fact is primarily not dependent on gender, but about who was considered as the most media literate (IE 1). Most of the time during forced migration the Internet connection was bad or even no Internet connection available. Also, asylum seekers had nearly no leisure time and therefore no time for entertainment. When arriving in a refugee camp, one usually had to recharge the mobile phone battery and had to recover first. The ones being on the run used their phone and free time to contact family or relatives (IE 1).

Some kinds of misunderstandings and rumors about benefits of being an asylum seeker in Germany were distributed through social media (IE 4) and these persisted when the asylum seekers attempted to come to Germany (IE 1).

> We found that many people thought that they would even get their own house as a refugee in Germany and that they would get lots of other social benefits. . . . Of course, there was always some truth to this kind of information. You will receive social benefits as a refugee and you will receive some sort of accommodation. . . . Later, a website on *Rumors about Germany* was created by the Federal Foreign Office for more accurate information about the benefits. And furthermore, videos about what really happens when you are a refugee in Germany were spread through different information channels. (IE 1)

But being an asylum seeker, there are also some stereotypes that one has to deal with in different host societies. For example, some people wonder why asylum seekers have bad clothing but also the newest version of an expensive smartphone (IE 1).

Written information is potentially a problem, and even more if not distributed in the mother tongue of asylum seekers. It took a long time for the German state and German official institutions to inform the people in their own language and to switch from written information to verbal or visual information, as this is preferred by asylum seekers.

> During the first rush of refugee applications, information was provided via posters with texts in various languages. The refugees were not aware of the posters as an information

> source and did not read them. They simply walked around and preferred to ask other people what they had to do and where they had to go and so on. So, it would probably have been easier to organize if you had put a few people in the middle of this place who could have passed on all the information verbally and repeated it 50 times a day. You would probably have had better results and a greater efficiency. (IE 1)

The fact that it is important to approach people in their mother tongue has been sometimes neglected by the German state and by German institutions for a few years. The idea was always that refugees should first learn the German language (IE 1). At first, it had to be recognized that refugees should also be reached out to in their native language. Suddenly, daily news and children's programs were produced in Arabic and efforts were made to reach asylum seekers through the information channels they primarily use. It is an important step to offer multilingual authorities but also different kinds of media in other languages to inform people (IE 1).

Especially official letters from the Federal Office and filling out administrative documents appeared to be an obstacle for refugees. As the documents are available in German, asylum seekers preferred to ask other people for help with filling in the forms and with translation of the official letters. Even for German native speakers it is sometimes difficult to understand the paperwork (IE 1). In the Netherlands, lawyers from the institution *Vluchtelingenwerk* help asylum seekers with paperwork. Because of information deficits, regular lawyers help out refugees with legal problems as well (IE 4).

The integration offices did not necessarily know the information needed by asylum seekers. The information has a lot to do with mastering everyday life in Germany and not necessarily world information. Things like "Don't touch a woman's head" are taught in integration courses and miss the point. The information needed for everyday life should be addressed and it should be taught what new arriving people really need (IE 1). Prof. Dr. Rianne Dekker explained that self-evident and common behavior is taught in the integration classes. For example, in the Netherlands, the integration courses teach that you have to congratulate women who have a newborn child. Her statement in this context was: "Shouldn't the Dutch people take those courses as well?" (IE 4) – As it is a kind of behavior that is welcomed and desirable by society but not every Dutch person behaves in this way either (IE 4). She highlighted, that especially the information needs of refugees should be considered for educational content and their knowledge gap adressed in integrational classes. Furthermore, in the Netherlands, asylum seekers have to pay for the courses themselves, which they are obligated to take. In addition, there is a great lack of information transparency about the quality of the courses and which ones are good or bad (IE 4).

Asylum seekers migrate during difficult circumstances (IE 4) and try to find a safe place to live somewhere to build a new home (IE 1). But the European government was sometimes reluctant to host refugees (IE 4). From the Netherland's federal office, official letters with asylum application information were spread through social media, which on the one hand served as an information update. But it was also considered a warning to not come to the Netherlands – asylum proceedings may take too long and one could be deported (IE 4). And further, deportation films were distributed through social media channels to threaten refugees with deportation (IE 1). All these methods somehow represent a totally misguided approach in relation to integration. "I think that misses the point. Most refugees are only looking for a safe place to live their life somewhere. For them it does not matter that the apartment will not be great . . . . It does not make sense to threaten them with deportation" (IE 1). Instead, social media should be used to inform asylum seekers about important aspects that will actually help them to get along in the new country, e.g., by letting authentic people who speak their native language and have successfully integrated talk about their background, their story, and their everyday life in the new country (IE 1).

12.2 Information behavior

At first, when asylum seekers arrived in Germany, they preferred to ask other people about most of the needed information. The common practice in Germany is to provide information to people in textual form. But, many of the people who seek asylum in Germany, firstly, do not know the German language and secondly, are often not used to receiving written information. They are more accustomed to visual or verbal information and usually prefer these (IE 1). For refugees, information from other people is considered as more important than online information and even considered as true, even if it could be half-truth (IE 2/3; IE 4). Furthermore, forums and social media are considered as an additional information source (IE 1; IE 4). They seek information not only from relatives and friends, but also from other people in their network and try to validate information with people they know and trust. However, information overload makes it difficult to decide whom to trust and what information is true (IE 4).

> We observed that people are trying to check the sources of the information – Where is it coming from? They check information with people they know and people they trust. So if they have someone from Vluchtelingenwerk or a family member or a neighbor they can ask, they will try to validate the information by asking them. They also will try and see if they find the information on different online sources and they try to compare information with their own experience. And, if information from a certain website or group proved to be true earlier, it is more likely to be trusted later on. (IE 4)

But it is difficult for people who are not part of the personal network to get in contact with asylum seekers. To contact them analogously by going into language classes or integration courses and talk to them in person has proven to be the most efficient way (IE 2/3).

Information needs

For various official offices in Germany it has been important to understand the information needs and information channels of asylum seekers to inform them properly. But, an even greater problem that has not been recognized is to inform asylum seekers in their native language. In Germany, the responsible administrations believed that learning the language first and then proceeding with further steps of the integration process is the most efficient way. But the information needs of asylum seekers were, and still are, about basic everyday living (IE 1). Further information needs include issues with paperwork and issues with the steps of the asylum procedure. “The problem is, you are at the authorities' offices and you do not understand these forms at all” (IE 1). As some of the documents of the authorities were only available in German, a Syrian asylum seeker himself developed a mobile application named *Bureaucrazy* to explain how to fill out the administrative documents. He has had the same information needs and prepared a solution for the knowledge gap of asylum seekers. Still, the most important information needs are related to the translation of administrative documents and authority letters (IE 1).

Other information needs of asylum applicants deal with basic needs of everyday life (e.g., taking the bus) or cultural difficulties and differences. World news and political information do not necessarily have priority. Asylum applicants have very direct and practical information needs and their information needs vary through different phases of immigration.

> You cannot work yet and at least in the Netherlands, you cannot start a Dutch language course before you are granted the residence permit. So then the needs are also to keep in touch with the people in the country of origin and, also, you try to get entertained. (IE 4)

First, they need information about handling everyday life in the new country. Especially, how to find a job with asylum status, or how to find and get a place to live. After asylum seekers are granted a permit to reside, the information needs are also shifting towards how to settle, how to integrate, or how to build a network (IE 4).

When it comes to seeking information with the help of digital media, first of all, one needs to have a digital device and access to the Internet or social media. Furthermore, it does not mean that asylum seekers know how to use the devices and media in an effective way to suit their information needs (IE 4). Most refugees have different information needs but not everyone has the skills and ability to inform themselves in a satisfactory manner (IE 2/3). To make it more simple to integrate and to tackle the demand of basic information, in Switzerland, the authorities prepare a welcome package for newly arriving asylum seekers with basic and everyday information including important addresses (e.g., integration offices, grocery stores, ambulances or doctors) and basic behavioral rules (e.g., traffic rules). Other countries should take this strategy as an example and prepare welcome packages for asylum seekers who are arriving in a new country (IE 2/3).

Information sources and channels

One of the most important questions is, how to reach asylum seekers. It has proven to be efficient to reach them analogously by visiting language courses and integration courses. Another way is to contact volunteers who help asylum seekers and arrange a meeting to speak with them in person. But, asylum seekers who proactively search for any needed information will usually be able to find it. Difficulties occur for people who do not search for the needed information (IE 2/3). To be able to inform asylum applicants and address their needs, it is important for the federal offices to know the used information channels. They want to reach and inform people and give accurate information, and to do so information channels that people are using to inform themselves need to be addressed (IE 1). In Germany, it is a common practice to distribute information in written format (e.g., letters from school for parents) but mostly, people from other cultures are used to other information channels, e.g., visual or verbal information from television or the radio. Being in another country one does not know the language and it is even more difficult to catch up with written information (IE 1). Therefore, many refugees use their smartphone to translate the texts with the help of online dictionaries or translation services (IE 2/3).

Hardly any of the asylum seekers did not have a mobile phone. The smartphone is not a luxury item for them, it is an everyday orientational tool, not only in the destination country but also during forced migration. During the escape, people were mostly using WhatsApp and the social media platforms that are based on their personal network (IE 1). For asylum seekers, especially people from their personal network are important information sources, as they do not know whom to trust. The communicated information from their personal ties are even more important and trustworthy to them than information from official sources and authorities (IE 2/3; IE 4). However, according to Prof. Carola Richter:

> But who can you trust? And what kind of communication or information channel can you trust? The majority told us that they rather trust German institutions and authorities. Especially if someone from the homeland . . . is also on the run . . . . For information and also for preparation, Deutsche Welle was quite frequently mentioned. Those people who were interested in the flight and had a feeling that they could go to Germany, already developed trust through the German information channels and Deutsche Welle. (IE 1)

Further, Prof. Dr. Rianne Dekker mentioned:

> Refugee migration differs from other types of migration by being very uncertain about your situation and the people are confronted with official information which they cannot always trust. . . . But then all the information that is circulating through social media is also uncertain, it is also rumor. So the key thing for refugees during this uncertain period of migration is finding out whom to trust. In some way it can sort of emancipate refugees, they have a lot more information than they used to have from traditional migrant networks, but at the same time, because there are so much information from so many different sources and you don't know the exact heritage of this information, it also becomes more difficult not to get the information but to find out what can be trusted. (IE 4)

In some countries, the official immigrant authorities and offices distribute information via social media (IE 1; IE 4). Social media is not only a source to get timely and practical information from the authorities and their social ties, but also additional information from other refugees (IE 1; IE 4). Asylum seekers use different information sources for different kinds of information, e.g., specific Facebook groups or forums which inform about the latest developments to access Europe through various routes and how to make use of smuggling services (IE 4). But people also exchange other information and ideas in online forums or Facebook groups and fan pages, e.g., exchanging recipes or cultural music (IE 1; IE 4). In many forums, information about Germany is written in Arabic (e.g., what to pay attention to when reading letters from authorities). There are also specific Facebook pages and groups for Syrians or other nationalities to exchange information about the country they are applying for asylum in (IE 4).

Another important media for refugees to get information is the television. A large market of TV channels in Turkish or Arabic is available in Germany and can be easily accessed via cable or satellite. Some of the channels report from the asylum seekers' home countries and about other world information. But, there are mostly channels without German interpretation and might be culturally influenced, therefore, the information about German news is missing. And, the German news broadcast *Die Tagesschau* is only available in the German language (IE 1).

Information and digital literacy

"One thing is, to have a digital device and to have access to the Internet and to social media, but it is also necessary to use it in a way that suits your needs and information needs" (IE 4). It is argued that some asylum seekers who come to Germany do not have the necessary digital skills to properly handle digital devices other than the smartphone (IE 2/3). The digital and information literacy competencies of asylum seekers seem to be sometimes poor and they express certain deficiencies. The digital literacy level of asylum seekers can be related to their educational level, financial means, and their language skills (IE 1; IE 2/3). Some adolescents with a refugee background rate their own skills better than they actually are. Whereby the younger people are more affine and confident in their own skills (IE 2/3).

Dr. Juliane Stiller and Dr. Violeta Trkulja developed workshops to teach asylum seekers and migrants digital and information literacy skills. They observed that the skills have to be taught in private lessons because so many questions are asked by the asylum applicants (IE 2/3). Another noticeable phenomenon is that most of the workshop participants are men. It is difficult to reach out to women because they take care of the children and do household chores (IE 1; IE 2/3). Furthermore, the attendees of the workshop are around 25 years to 40 years old. For the ones older than 40 years, it is hard to learn a new language, and on the other hand, also difficult to learn how to handle digital devices, e.g., a smartphone or a computer. Those who are digital natives almost take digital skills for granted as they grew up with a smartphone, but older people are not familiar with it (IE 2/3).

Other than that, the asylum applicants did not have the need for a stationary computer or laptop in their former home country. Especially people from Eritrea or other African countries mainly used their smartphone and did not have other digital devices. And, in Germany, they still do not have the need for other devices and therefore do not have the necessary devices and equipment to learn digital literacy skills (IE 2/3). Direct information seeking is not a common activity for some asylum seekers. Due to language difficulties and lack of basic digital literacy skills, they have problems with information research. They are only able to use search engines and job searching engines rudimentarily. Procedures for job applications are different – and having a job is equal to affording a living. Also, some countries educate differently and with other standards than, for example, Germany (IE 2/3).

When speaking of information literacy skills, it is important to consider information validation. Different strategies to validate information were acquired by asylum applicants (in order to be able to detect fake news and rumors on the web and on social media). They are aware of the fact that not everything can be trusted. A common strategy is first, to try to find and check the original source of the information and compare it to other online or offline sources. Afterwards, people they know and trust are asked, i.e., other refugees, people from official organizations, or neighbors. The information is also compared to their own experiences. If a website could be trusted earlier, they are more likely to trust the source again (IE 4).

12.3 ICT and media

In general, some asylum seekers do not have the need for other digital devices than a smartphone to access the Internet. With the smartphone, one can do nearly everything, especially if it is Internet-related, no longer needed are stationary computers, laptops, or tablets (IE 1; IE 2/3). For people in Germany, the smartphone is primarily seen as a luxury and entertainment provider, but for asylum seekers it is mainly a multifunctional and orientational tool. Pictures can be taken, money transferred, and people contacted. During the escape, it is an essential tool to plan one's route, e.g., to find the best way through a fence via Google Maps (IE 1). Also, documents and texts can be translated and needed information searched for. Furthermore, the smartphone is used to reach family members and friends who they had to leave behind in the former home country or who live in a different country (IE 1). But, of course, they also use smartphones in their destination country to get entertained by watching YouTube videos or listening to music. Asylum seekers mainly use the Internet to socialize, contact other people, or to get entertained in some way. Information seeking is not a common activity of some asylum seekers and search engines are only used rudimentary (IE 2/3).

Nearly everyone who fled their former home country has a smartphone or mobile phone, female forced migrants and older generations as well. They also mostly use it to communicate with relatives or friends, and much less to do research (IE 1). In Germany, like in the home country, there is no need for other digital devices (IE 2/3). The ones who had a more advanced educational background were more likely to be familiar with other devices and felt more comfortable with using the Internet (IE 1). It is important to possess digital devices and to have access to the Internet, but another important point is to use them in an effective way to suit your needs (IE 4).

There is a large market for apps that were especially developed for asylum seekers, but before they will find an app that is specific for them, they will more likely use a more general app or platform for the same benefit (IE 4). A platform for matching asylum seekers with a workplace was developed and released, but people with a refugee background did not use this platform. For people in

Germany, a job is seen as some kind of self-realization, whereas in countries of many refugees it is mostly said that jobs are taken to afford a living. It exists a different understanding and meaning of the concept of "having a job" (IE 2/3).

Other than that, there is a wide range of TV channels available in Arabic or Turkish that can be accessed via satellite or cable connection. The channels are widely available and in the native language of most asylum applicants, therefore, people are able to watch all kinds of informational or entertaining channels in their mother tongue. When it comes to news services, there is a split world of information. The channels report a lot about the home countries but not especially about Germany. Information from the German interpretation, e.g., Die Tagesschau, is missing. And also, the essential and necessary information on how to master daily life in Germany is not available in video formats. It is an important aspect for the integration process to offer those formats as well (IE 1).

12.4 Social media use

Social media plays an important role for asylum seekers as it is a source of information and it makes it possible to stay connected and to get in touch with relatives and friends (IE 1; IE 2/3; IE 4). As in the beginning, after arriving in a new home country, most of the asylum seekers are not allowed to work, they try to get entertained in some way. Watching YouTube videos is a common activity, as it is possible to watch videos in their mother tongue. Social media services and messengers like WhatsApp, Facebook, and Skype are used by asylum seekers to communicate and get information from other refugees (IE 2/3; IE 4). Whereby the platforms used by the refugees mostly depend on their personal network. And, during the escape, asylum seekers often expand their personal networks, they meet people who have similar experiences, e.g., traveling the same routes or finding refuge in the same city (IE 4).

Facebook groups and Facebook fan pages as well as forums are used to exchange information with others, especially about mastering everyday life in the new country, but not so much about news (IE 1; IE 4). For example, recipes and cultural music are posted and discussed in Facebook groups and on Facebook pages. Furthermore, in forums, the latest developments to access Europe through escape routes and the use of services of smugglers are discussed (IE 4). Some people feel more confident in dealing with information from social media than others. An important aspect to consider is the distribution of fake news and misinformation on social media. Different social media have different limitations and most people, especially being an asylum seeker, are aware of the need to validate the information they receive online. There are different strategies to validate social media information and further online information. One of them is by checking the source of the information and where it originates from (IE 4). "The asylum seekers will try and see if the information can be found at different online sources and furthermore, they will try to compare the information with their own experience, so if information from a certain website or group proved to be true earlier, it is more likely to be trusted later on" (IE 4). Or they check information with other people they know and they can trust. That might be someone from Vluchtelingenwerk or a family member or a neighbor they can ask. "So they [refugees] have different strategies which I think are not really unique to refugees, and not present among all. But you need to be aware of different limitations of social media and most people are aware that not everything can be trusted and they will try to find ways to validate information" (IE 4).

When it is decided to leave the former home country in order to search for asylum elsewhere, the Internet and social media are not the only sources to be relied on, because there are still ways to escape and to get in contact with migrant networks or persons from personal networks who are able to help to escape (IE 4). Prof. Dr. Rianne Dekker states:

> I believe that migrant networks, so personal networks are also still there and independent of social media. Further, I believe if you are not an Internet user you will still have possibilities to be in contact with family members or other people and who can help you out to migrate. But I do think that it makes the life for some refugees easier to be able to explore new contacts who are not yet in their personal networks and are valuable, because they have experienced similar and they made similar steps in migration. Or maybe they chose the same routes or they ended up in the same cities you were interested in. (IE 4)

12.5 Age- and gender-dependent observations

During the flight, opinion leaders were always men, but this was not necessarily a question of gender, but rather of being media literate (IE 1).

> So I think you really have to state that the group that came here is essentially a lot of young men or unaccompanied minors, and also a majority of families that are in the 30 to 50 age range. And furthermore, there are also not many digital natives among the women and I think in this group it is indeed the case that only certain types of communication are made and not this intensive research. (IE 1)

Women have to be considered much more when it comes to asylum-related information distribution, as they take care of the children and therefore do not have the time to search for, or to take, a job (IE 2/3). A distribution of roles was observed, whereby the women have to care for the family. Women often did not have the time or pushed the men forward to answer questions:

> Women are often managing the families and picking up the children, we noticed this ourselves in our study. We had always tried to ask a representative amount of women, which at least is shown in the statistics of those arriving here, but that was very difficult because in the family groups . . . the men were often pushed forward as the supposedly knowledgeable ones or those who have the time to answer our questions. (IE 1)

The study of Prof. Dr. Carola Richter shows that nearly all women possessed a smartphone. They use the smartphone primarily to communicate with relatives and to stay in touch with friends, while they do not use it much to do some research. But, in general, there was no evidence for a different information behavior of men and women (IE 1).

Younger people are more affine with the media and more confident about their digital skills. Generally, in digital literacy workshops as far as Internet activity is concerned, the youngest are around 25 years old and the oldest are around 40 years old. After the age of 40, it is very hard and very tedious to learn a second language (IE 2/3). There is an overall social problem with a lack of digital literacy skills. This is partly related to the level of education or financial means and general language skills (IE 1; IE 2/3). Furthermore, those who fled their former home country as a family mostly had a better educational level and were more familiar with the Internet (IE 1).

12.6 Conclusion

When arriving in a new country, asylum seekers are first and foremost in need of information about mastering everyday life. The lack of language skills makes it even more difficult to find one's way around in a foreign country. A large amount of information in Germany is distributed in written form. The information needs of asylum seekers differ in various phases of migration and integration. First of all, asylum seekers have difficulties with understanding official letters and documents from the offices and how to fill in official and administrative forms. For this, they preferred to ask for other people's help. In the Netherlands, lawyers help asylum seekers with those documents. Furthermore, for other information, asylum seekers prefer to ask other people from their personal network or neighbors and people from the offices and even trust them more than other information sources. Information from the Internet is not always trusted, as there is a chance for potential misinformation. Asylum seekers are aware that they have to validate information from the Internet and try to check the original source or other sources for the information. Further, hereby they prefer to ask other people as well. And, sources that have been proven to be true earlier are more likely to be trusted again.

Whether refugees own a cell phone or smartphone is mainly based on their financial means. Most of the asylum seekers do not have digital devices other than a smartphone, as they do not have a need for them. Therefore, the skills to handle these devices are sometimes poorly developed. For asylum seekers, the smartphone is a very important tool as it acts as a computer that can be carried around and does not need much space. Even during escape, they are able to use it for important activities, e.g., planning the next steps. It does not only serve as a multifunctional tool during escape, it further helps them with orientation in the host society and is often used to help with translation. In order to stay in contact with family and friends, especially those who are left behind in their home country or found refuge in another country, social media services are used. Instant messaging services like WhatsApp and Facebook Messenger were often mentioned. Some even use Skype for video calls. Asylum seekers also watch videos on YouTube and listen to music to get entertained. These activities are also carried out with the smartphone. Television channels broadcasting in Turkish and Arabic are also available in Germany. But only a few make use of it.

A few age-dependent and gender-dependent differences could be detected. There is a distribution of, to be considered, more traditional roles if family life is concerned. Some of the women take care of the children and are responsible for the household, while men will try to get a job to make a living for the family. But there was no evidence for differences in media usage and information research. For the age related findings, similarities to people who are not asylum seekers can be noticed. The digital natives who grew up with the smartphone and other digital devices take digital skills for granted. For people who are older than 40 years, it is very difficult and hard to catch up with digital skills. Furthermore, for them, it is even more difficult to learn the new language of the host society.

In conclusion, asylum seekers face different challenges when coming to a new home country. They need to learn a new language, need support with paperwork, and do not know how to get along with everyday life and cultural differences at first. However, using their smartphone and different social media platforms to stay in contact with family and friends and using forums or Facebook to ask questions is seen as being helpful.

13 Discussion

In its empirical parts, this work is an attempt to investigate the information behavior and media practices of asylum seekers after arriving in a new country. Insights could be gained by applying a multiple methods approach. Different ICT and media forms relevant for this study are traditional media such as print media like newspapers and magazines, books, but also broadcast television, the radio, letters, phone calls for a landline, and any other forms not connected to the Internet. New media includes all digital, interactive, and machine-readable media. This is digital media such as software, video games, web pages, databases, digital audio, digital documents and books, and social media.

For the theoretical framework, the theory of information behavior (Wilson, 2000), i.e. information seeking and information reception (Savolainen, 2007; Spink & Cole, 2004), and the information horizon (Sonnenwald, 2005) as well as the Uses and Gratifications Theory (U>) (Katz et al., 1973; McQuail, 1983; Shao, 2009) were applied. Derived from this theoretical framework, an intuitive research model was created, serving as a basis for this work. The model describes the information horizon related to media and ICT usage of asylum seekers. Information is exchanged with other people, based on different needs, and via various forms of media, such as online and social media. Information is produced in this process, especially on social media and online media. Based on this, it can be assumed that the results correspond with the proposed model. However, it needs to be recognized that various precautions were taken by asylum seekers when producing information or online media products such as videos. Some explained how they do not easily trust the information found on social media and are cautious when showing their faces on social media. In contrast, other participants share details about themselves and their families more freely.

The present work is a unification of multiple studies which can also stand alone and answer different research questions which together contribute to the answer of how the behavior of asylum seekers changes in the target country and which information needs they have. In the following chapter, the main findings of the research questions are summarized and discussed. Finally, recommendations and a brief conclusion as well as limitations of each investigation and an outlook are presented.

https://doi.org/10.1515/9783110672022-013

13.1 ICT and media practices for integration – A literature review

Looking at the first research question "How do asylum seekers use ICT, online and traditional media as well as social media for integration according to the literature?", key themes identified in the literature review are the use of smartphones to 1) find information, 2) communicate with friends and family abroad, 3) meet locals, 4) meet peers, and 5) counteract boredom (Merz et al., 2018). The empirical results of the study at hand support these main themes. The U> served as a basis for the investigation and all asylum seekers expressed different needs concerning these motivations. Information can be acquired about different aspects of life, such as knowledge about the new home country. Existing personal contacts can be maintained and new ones formed, entertainment-needs be satisfied, and things about daily life and oneself be shared. This desire for self-presentation was also investigated by Leurs (2017). The literature review revealed contrasting results concerning ICT ownership and media usage. Some suggest there are no divides in smartphone ownership (Merisalo & Jauhiainen, 2020), others explain there are differences (Ahmad, 2020; McCaffrey & Taha, 2019). According to the empirical results of the study at hand, nearly all participants own a smartphone and have access to various ICTs and media.

It was revealed how first contact points for information sharing seems to be human rather than technological (e.g., Alencar & Tsagkroni, 2019; Bletscher, 2020). Our results and interviews indicate that, indeed, social contacts are important for information. Furthermore, participants expressed how they ask teachers, other asylum seekers, family members, or social workers about information. However, the role of ICT should not be underestimated in this regard, as not all forms of information are shared face-to-face. Women prefer to ask about information via WhatsApp. Some information, such as the whereabouts of the family, can only be exchanged by online services or ICT, if the family lives abroad, for example. The most pronounced information needs expressed by asylum seekers according to the literature are about language learning, education and employment, health, law, news, housing, and everyday information. Our study also revealed language, health, and news to be of most importance.

Regarding the popularity of different services, WhatsApp and other messaging services were the most important for communication purposes, Facebook to find information, and YouTube for entertainment. Twitter and Snapchat are barely used. These findings are similar to our studies' results. However, YouTube also has a high status for information purposes, such as language learning. Considering apps, most helpful are social media apps, translation services, apps for completion of school work, and albeit not considered for our study, e-payment apps, and location finding apps (Coles-Kemp et al., 2018).

Some problems for asylum seekers in relation to ICT and media usage were observed according to the literature review and our results. The literature review shed light on fears concerning media usage, for example, parents are worried that they are not able to help and protect their children when it comes to, e.g., Internet usage (Coles-Kemp et al., 2018). This notion seems to be expressed by our adult female participants as well, or at least some perceive their media skills to be rather low.

Furthermore, according to studies from 2015 and 2017, finding information online is not familiar territory to asylum seekers (Mikal & Woodfield, 2015; Tirosh & Schejter, 2017). Also, it could be problematic to acquire fluency in a foreign language by relying on ICT for translations. However, all of our participants, adults and children alike, repeatedly explained how they use specific services such as YouTube to learn German. Language learning apps are used by more asylum seekers since living in Germany. They also stated how they are able to find information by using search engines and other platforms. Therefore, ICTs and media can also be beneficial if used to acquire the language.

Some argue that by forming ties with others and being able to speak the new country's language, the reliance on social media weakens (Köster et al., 2018). Or explained the other way around, by relying on social media for social contacts, integration could be difficult (Marlowe, 2020). This sentiment was shared by our participants. Some explained how hard it is for them to form relationships with Germans due to the reserved nature of German people (IP13, IP25). However, others explained how they have German friends and contacts who are eager to help, such as neighbors (IP5, IP11, IP14). By not being able to relate to the people in the home country, focusing on online contacts or already established contacts is likely. Since social ties can be formed via social media, the situation may not be as dire, making social media a double-edged sword in this regard. Focusing on the information behavior regarding the new living conditions of asylum seekers after arriving in a new home country, media and ICT usage was considered as part of a new life.

13.2 Identifying ICT and media practices in a new country – Age- and gender-dependent results

In total, three research questions were investigated as part of the study on how asylum seekers use ICT and media in a new home country, in this case, Germany. While the results for the research questions cannot represent all asylum seekers or even be sufficiently answered, some insights could be made as a momentarily description of the asylum seekers' lives. In the following, what ICT, online and traditional media as well as social media asylum seekers use and if age- and gender-dependent results can be observed (RQ2a) will be discussed. Concerning the overall usage distribution of various media forms and ICT, the most popular ICT are smartphones, the Internet, and the TV among all study participants. Boys and men seem to have the biggest interest in technological applications, as more of them use all of the ICTs proposed by the survey in contrast to the other study groups. Women seem to be more focused on print media, as almost all of them read books and newspapers but show low percentages for all ICT categories except smartphones, the Internet, and TV. For the girls, the distribution is a little lower for ICTs, but still comparatively high, and all of them like to read books. The distribution of the usage percentages seems to be reflected in how asylum seekers perceive their skills to accurately use ICTs (RQ2b). On a five-point Likert-scale, with 1) meaning "very bad," 3) "neither bad nor good," or 5) "very good" skills, the perceived ability is rated the highest by the boys and girls with a median of 4 and an IQR of 1.5 and outliers at 3. Men also rate their abilities highly – with a median at 4, and IQR of 2. Women are the least confident in their abilities, with a median at 3, an IQR of 2, and outlier at 2. This discrepancy between confidence-levels of the participants could be traced to the fact that there is no time to use media, especially for parents, as they have to look after their children in addition to learning a new language and adapting to a new culture, as was expressed by the studies participants. Furthermore, some of the women seem to trust their husbands when it comes to information practices. This was explained by one woman during the interview rounds: "My husband organizes a lot of things concerning education, legal aspects, health, and news because he is more confident in speaking German" (IP14).

These findings are mirrored in the application of ICT, online and traditional media as well as social media of asylum seekers to satisfy their needs for information, entertainment, social interaction, and self-presentation, and the corresponding age- and gender-dependent results (RQ2c). Looking at the needed information by the different groups, adults are interested in all categories. In general, all information categories were ranked rather high, except for employment and religion for women, and religion for men. The rather low interest in employment could be related to the fact that all of the participants had to finish the language courses before being employed in Germany, making it less of a priority at the moment. The interview rounds also suggested that the men are more likely to be the ones working or searching for education online while the women look after the children. This is also confirmed by the literature (Merisalo & Jauhiainen, 2020) and our expert interviews. It was stated how some of the participants are no longer interested in religion after arriving in Germany. Information about the home country was either sought after or completely avoided. According to the literature, important information is related to legal aspects, education and profession, accommodation, and family (Schreieck et al., 2017).

Important for women are information related to health, especially for their children. Information about Germany and general news are important as well, highlighting the desire to learn about the new home country and its different aspects such as culture, customs, and events. For men, most important are information related to education, news, and information about Germany. Naturally, as the children were rather young, there was no interest in many of the information categories. They mainly need information about homework and news. However, one of the older teenagers was interested in employment. Regarding the use of online and print media in correlation to information needs, the young generation is able to collect needed information via online media. The children mostly use search engines, and also WhatsApp and YouTube. The other services are not utilized. For women, there is more emphasis on social interaction to gather information, such as messaging services like WhatsApp. They also read print media such as newspapers more than men. In contrast, men are focused on information acquisition by using search engines, YouTube, and the TV (albeit, women are also interested in using search engines and YouTube, however, to a smaller extent). A man expressed how he does not trust information circulating on social media such as Facebook as he is highly aware how misinformation is shared on these platforms (extensively discussed in the chapter (chapter 8)). Therefore, the information and media literacy skills and precautions are also relevant in the new home country. All of this suggests that information needs and information behavior changed due to experiences of the study participants.

The entertainment and socialization dimension are also seen as highly important. However, it was also expressed by the adult participants that a lot of applications for social media and entertainment were deleted and not used anymore after arriving in Germany. Regarding the children, for entertainment, only YouTube, apps on the smartphone, and additionally for the girls TikTok are relevant. The adult participants, if they have time, like to use YouTube; men also use Facebook, the women focus more on WhatsApp. Almost all socializing needs in context with digital applications are satisfied by using WhatsApp; women also use Facebook. Services that are not as relevant anymore, such as Skype and SMS, are seldom used by the younger participants. It was expressed how these services are only used if there is no alternative (such as no mobile data on the smartphone). Women however seem to write SMS quite a lot. Considering self-presentation, it was obvious how some of the participants are not interested in it or only share pictures about their children, for example, on

Facebook. The children expressed how they like to make videos for TikTok but do not show their faces. Others explained how they only want to share details with friends. All of the girls however like to share information about themselves on various media, but mostly WhatsApp; men and women use Facebook the most for this.

13.3 Home country and now – Comparing adapted media and ICT practices

In order to be able to inform asylum seekers through the appropriate channels, it is important to understand which channels they use. In this regard, it is also necessary to know which channels they are already used to, since they have applied them in their former home country, and which online and traditional media, social media services, and ICTs they started to use since living in the new home country, Germany (RQ3a). Furthermore, to evaluate the commitment and engagement toward a service, media, or device, it is not only important to know which channels are used and were used by asylum seekers, but also how frequently (RQ3b) and what the motives are to apply certain ICT, social media, or online and traditional media (RQ3c). Insights on this were obtained with the help of a survey. All participants were adults between 23 and 54 years old, therefore details and results about the behavior of children and the elderly are missing.

Starting with the results related to the question on what ICT, online and traditional media as well as social media asylum seekers used in their home country and in Germany (RQ3a), some differences were observed. Many asylum seekers have already used various social media and online media in their home country. However, several more asylees started to use some of them after their arrival in Germany. YouTube, Facebook, messaging services, and Instagram were especially important social media in the lives of asylum seekers before they had to leave their home country. Even more asylum seekers started to use them in the target country, i.e. Germany. Also, some social media services that were only used by a few asylum seekers in their former homeland, speaking of Snapchat, TikTok, and LinkedIn, are now used by some more asylum seekers since living in Germany. Skype is the online media used by less asylum seekers since living in Germany, which was also confirmed by the conducted interviews. Taking a look at online media, translation services in particular are now being used by much more asylum seekers since living in Germany. In comparison to the asylum seekers' former homeland, more asylees write e-mails since living in Germany as well. Further online media which more asylum seekers started to use in Germany are streaming services, e.g., Netflix and also language learning apps.

For traditional media, asylum seekers started to write letters more often. As also obtained by the conducted interviews, a few participants do not write SMS since living in Germany. For other traditional media, i.e., books and newspapers or magazines, and also for ICTs (smartphone, television, computer, notebook or laptop, and tablet) the amount of asylum seekers who use them since they are in Germany is nearly the same. The only ICT which is no longer used by the majority of asylum seekers in Germany is the landline telephone.

Coming to conclusions regarding the question on how frequently asylum seekers use or have used ICT, online and traditional media as well as social media in their former home country and in Germany (RQ3b), various social media services and online media are used daily or a few times a week by most of the asylum seekers. Especially translation services and e-mail became very important online media in asylum seekers' life since living in Germany. Translation services seem to be a great support for asylum seekers in the new country regarding the language barriers and most of them take advantage of it on a daily basis in Germany. Most write e-mails a few times a week or daily in Germany, while in the former home country most only did so a few times a month. Podcasts and especially streaming services and language learning apps are used much more frequently by asylum seekers since living in Germany. The usage frequency of search engines, news websites, and encyclopedias are nearly the same for the homeland and for Germany. Only digital games are played less often by asylum seekers since living in Germany. Regarding social media, the majority of asylum seekers use messaging services, Instagram, and YouTube on a daily basis in Germany. YouTube was the only service used less frequently by some asylum seekers in their former homeland. The social networking service Facebook and micro-blogging service Twitter are used daily or a few times a week by most of the respondents in Germany and were used equally often in the former home country. Live streaming services are used much more since asylum seekers live in Germany, half of them do so on a daily basis. Not many asylum seekers have used TikTok in their former homeland. Since living in Germany, more asylum seekers began to use TikTok, most of them do it daily or at least once a week. Regarding the services, only Skype is used less frequently by asylum seekers since living in Germany. Skype's main function, taking video calls and chatting, is integrated in messaging services as well and these services therefore replace Skype in some way. As already reported by other investigations, the smartphone is the most used ICT by asylum seekers (e.g., Emmer et al., 2016; Coles-Kemp et al., 2018). The survey results also confirm this: The smartphone is the only device used daily by all respondents in Germany and was also used daily in the former homeland. The most noticeable changes can be seen for the landline telephone. Although the majority used the landline telephone daily or a few times a week in their home country, it is used by much less asylum seekers and less frequently since living in Germany. Here, the smartphone replaces the landline telephone. Moving on, the computer, laptop, or notebook and the television were used daily by asylum seekers in their homeland, in Germany, most still use them daily or a few times a week. Furthermore, more asylum seekers use the tablet in Germany on a daily basis than in their former home country. Regarding traditional media, letters are written more often by asylum seekers since they live in Germany. One possible explanation would be that communication with German authorities is mostly done via letters. Books were read daily by the majority in the home country, in Germany most do it daily or a few times a week. The same results apply to writing SMS. While in the home country, most read newspapers and magazines on a daily basis or a few times a week, most do it less frequently in Germany.

Taking a look at the question on what asylum seekers' motives in their former home country and in Germany were to apply ICTs, online and traditional media as well as social media (RQ3c), only a few changes were detected. In both the asylum seekers' home country and in Germany, Facebook was used and is used for information and socializing. In their homeland, Facebook was further used for entertainment and self-presentation purposes. The usage of Instagram, Snapchat, and LinkedIn is and was motivated by information and socializing, while Instagram is and was used for information, socializing, and entertainment as well. Messaging services are used for all four motives, but especially for socializing. Regarding the usage of Skype, it was used for socializing in the asylum seekers' homeland. In Germany, most now have a neutral opinion on this. Video-based platforms, speaking of YouTube, live streaming services, and TikTok, and also the image-sharing service Pinterest were and are mainly used for information and entertainment. Information was and is the main motive for asylum seekers to use the micro-blogging service Twitter. Dating apps are used in Germany for entertainment, socializing, and self-presentation. Coming to ICTs, the smartphone is and was used for all four motives in the asylum seekers' home country and in Germany, but mainly for information and socializing. The computer, laptop, or notebook is especially used for information, entertainment, and socializing purposes. Information, entertainment and building a personal identity were motives to watch television, in the asylum seekers' homeland and in Germany alike. The tablet is and was used for receiving information and being entertained in the former home country and in Germany. In the home country, the tablet was further used for self-presentation. Regarding the motives of asylum seekers to use traditional media, the most noticeable shift was reported for the landline telephone. While asylum seekers used it for all four motives in their homeland, they do not agree to use it for any of the motives in Germany. Here, the smartphone seems to replace the landline telephone. In contrast, while in the asylum seekers' former homeland, letters were not written for any of the four motives, in Germany many agree to write letters for information. All four motives apply for reading books and writing SMS. Books are especially read for information, personal identity, and socializing (meaning talking with others about the read information), and SMS are especially written for information and socializing purposes. Newspapers and magazines were read for all four motives in the home country, but especially for information. In Germany, asylum seekers only read newspapers and magazines for information. Regarding the motives for online media, participants were not asked for motives to use translation services and search engines as these services are mainly used for information (i.e. translation and searching). E-mails were and are written by asylum seekers for information and socializing, but especially for information purposes since living in Germany. Asylum seekers listen to podcasts for information purposes and furthermore to build their personal identity in Germany. In the homeland, podcasts were mainly listened to for information and entertainment. Encyclopedias are mainly read to receive information, but also for socializing. Streaming services, learning apps, and news websites are also used for all four motives, but here the main motive for streaming services is entertainment, and for learning apps and news websites it is information. Furthermore, many used news websites for socializing (communicating) purposes in their homeland. Finally, asylum seekers play digital games for entertainment, even more since living in Germany.

13.4 Information needs of asylum seekers in a new country

Moving on to the question on what kind of information asylum seekers search for in online forums (RQ4), a content analysis of the forum *Wefugees* helps to gain insight. The category "Other Questions" was investigated and resulted in 23 categories and 139 analyzed questions. The posts often concerned questions about living with children in Germany and how to receive a residence permit. The most asked about category is related to documents and legal status of the asylum process. This category is split in several sub-categories, so a high number of questions was expected. As the results of our interviews display, letters from authorities, especially of legal matters, are difficult to understand, so asking about them in forums is not a surprise. Ranking next are questions about employment, media, and marriage. Employment often focused on issues related to the asylum status, for example, "Can rejected asylum seekers work in Germany?", "I have a *Duldung* status – Am I allowed to have a trading license?" Asked about often is the topic of work in relation to education certificates and documents and how they could be verified. The media aspect was the third important theme. The questions revolved around ICT such as mobile phones, the computer, and online media such as YouTube, websites, or apps. For example, a need for translations of German websites was expressed: "Where can I find English or Arabic translations or translators? I do not speak German and all refugee help websites are in German." Others wanted to know where to buy a cheap mobile phone. Concerning marriage, the questions revolved around documents. Often, the question regarded marrying another person who is not an asylum seeker or whose application was declined. "I am a refugee in Munich, Germany. I have a three year residence permit in Germany. I want to marry my girlfriend, but she is a refugee without documents and she lives in a French refugee asylum home. How can I marry my girlfriend so we both can live together in Germany?" Other important factors are language, travel, and institutions. Not often asked about was information regarding family reunification, the interview of the Refugee Status Determination process, and culture. Related to the findings regarding asylum seekers' information needs, some mobile applications were already developed. *The Migration Advice Service for Adult Immigrants* (MBE) initiated a project – the "mobile messenger-based advice service app" (Mbeon, n.d.) and information platform *mbeon*. The platform provides information about various topics, including learning German, work and careers, health, housing, family, and residencies and furthermore gives advice via chat on topical questions from individuals. It is another step (similar to other initiatives described in chapter 4 (integration initiatives)) towards helping asylum seekers to integrate into a new society and to find their way in the new country.

13.5 Experts' insights – Difficulties and information practices of asylum seekers

Last but not least, looking at the insights made by experts regarding the information behavior as well as ICT and media usage of asylum seekers (RQ5), we wanted to broaden our understanding about asylum seekers' information behavior and usage of media. Four experts, three from Germany and one from the Netherlands, were interviewed on their work with asylum seekers' media practices. Many of the things discussed during the interviews are similar to the empirical results we were able to gather. Learning a new language, using different social media to stay in contact with friends and family, and using Facebook groups to ask questions about the integration process, and specific challenges related to integration were mentioned.

During the interviews, it was explained that asylum seekers often only own a smartphone and no other ICT such as a PC for financial reasons. Therefore, it was argued that not all asylum seekers possess highly developed digital skills. As we did not ask for specific skills, we only focused on the perceived skills which were, except for the women, perceived as rather positive; there can of course be a discrepancy between those two. Accordingly, it was mentioned how these skills are sometimes perceived as being better than they actually are. The literature also discussed how the selected leader of a group during forced migration was the one with the best media skills. Of course, the younger ones who grew up with media are often also the ones with the best skills. Therefore, it can be assumed that there exists a mixture of highly skilled asylum seekers and some with less digital skills.

It was also described that, apparently, information in Germany is distributed only in written form and mostly in German. According to BAMF, all asylum seekers receive important documents in their native language. However, different forms of media could be helpful to reach asylum seekers, such as YouTube videos in different languages. This would be helpful since understanding official documents and letters is difficult enough, especially when learning a new language. It was also mentioned how lawyers help asylum seekers to understand those letters in the Netherlands. It was obvious how asylum seekers do not trust all information circulating on the Internet or social media and therefore want to validate information, mostly by relying on people from their personal networks such as people from official offices or neighbors.

Further, specific family dynamics were discussed. Men seem to be the ones wanting to be employed while the women take care of the children. It was also explained how men were mostly the ones being interviewed as women refrained from this. Men were also sometimes presented by their wives as being more knowledgeable (mainly because of language skills). All in all, the results of the expert interviews mirrored some of the other reported findings.

13.6 Recommendations

Based on the results of this work and previous studies, a few recommendations for different stakeholders are suggested. Migration authorities already provide various initiatives to help asylum seekers with orientation in the new country and furthermore with migration as well as integration processes. Offered language courses and integration courses are a first step toward this. Further initiatives from other organizations are also present (e.g., Wefugees, mbeon, HELP@APP). Somehow these initiatives are not forwarded to or not known by asylum seekers as none of the asylum seekers mentioned any of the initiatives during the interviews. Here, an overview website or information sheet as well as a better communicative transfer of these initiatives could be considered by authorities. Also, a helpful tool for this and further information exchange could be one central information system with updates about new initiatives and further developments or changes. Additionally, information sent out via the most used channels, speaking of SMS, WhatsApp, e-mail, and letters (asylum seekers can decide which information channel they prefer and whether they want to receive the information), similar to a quarterly newsletter would benefit the communication of latest information. Many asylum seekers who took part in the interviews were reading the free newspaper handed out by some cities, a yearly newspaper in different languages about developments may serve as another option.

Information literacy plays an important role to take part in a society and further to be able to find information for orientation in a new society (Lloyd et al., 2013). Therefore, our recommendation is to offer information literacy courses, especially for older and less information and media literate asylum seekers, in line with the integration courses. This will allow them to take advantage of various useful tools (e.g., translation services, search engines, communication channels) and other applications and initiatives and to be able to find information more efficiently.

In Switzerland, the government already provides some kind of welcome package with key information about everyday life. One suggestion is to provide welcome packages for newly arriving asylum seekers in other countries as well.

Another important aspect are letters from official authorities which are sometimes hard to understand. Initiatives like the app bureaucrazy developed by one affected asylum seeker could be extremely helpful to fill out those letters and forms. Further, it can be highlighted which steps of the asylum process are needed to take next.

In the work at hand, the role of German citizens was only considered by asking asylum seekers if and how often they communicate with German citizens and, furthermore, whether they would like to have more contact and social interaction with German citizens. Indeed, the majority expresses the desire to be more involved in the everyday life of them but mention the language barrier as a difficulty. Therefore, some do not seek contact with others who do not speak the same language. A recommendation for German citizens is to proactively engage with asylum seekers and, if possible, to volunteer. They further may start to get in contact with neighbors who are (forced) migrants. Additionally, some universities and other institutions could offer some kind of "language partnership" or "buddy program" in which one could participate. As it was observed how children pick up the new language much faster than the parents and form friends easily in school, parents could plan meet-ups this way to foster social interactions.

13.7 Conclusion

Coming back to the subtitle of this monograph, "New Life – New Information?" and the main purpose of this research: to investigate the information behavior as well as the social media, offline and online media, and ICT practices of asylum seekers and refugees in a new country. Indeed, the asylum seekers' information behavior changed as they have a great need for various information after arriving in a, for many of them, new country. They are not familiar with the foreign language, have to adapt to a society and new circumstances and have to undergo the country's asylum procedure. They furthermore have to comply with juridical regulations made by the government of the country, e.g., in order to be able to be employed. The needed information by asylum seekers are mainly addressed to everyday life in the new country, official documents, and legal aspects, especially related to the asylum procedure.

Almost all asylum seekers own smartphones and use them daily. It is often referred to as a multifunctional tool. The smartphone provides access to the Internet and thus, various social media as well as online media which are used for information (knowledge), entertainment, and socializing. Needed information is sought via many of these services, e.g., YouTube, encyclopedia, and news websites. Particular questions are asked in online forums and Facebook groups and, besides, it is preferred to ask people from personal networks or teachers and social workers. Some asylum seekers are aware of circulating misinformation in online environments and therefore can develop trust issues regarding unfamiliar and new online information sources; they often rely on sources approved to be true in the past.

Messaging services make it possible to stay in contact with relatives and friends, not only those in the country of origin but also with the ones who were met during the journey or live in other countries and other parts of Germany. The landline telephone and Skype are no longer used for this purpose by many. Learning a foreign language is a priority for most asylum seekers, as they expect and hope to be granted a residence permit and further want to build their life in the new country. Many asylum seekers therefore started to use language learning apps since living in Germany and also started to use translation services – most use them on a daily basis – to overcome prevailing language barriers. Nevertheless, a language barrier can make it difficult for asylum seekers to establish new social contacts with citizens from the new home country, although many report they would like to.

Various ICTs, online media, social media, and traditional media further help asylum seekers to tackle boredom, even though some adults report they have to learn the foreign language first, try to find employment, and have to take care of the children and do household chores, and therefore do not have much time for entertaining activities. Children of asylum seekers use online media mainly for school and homework and social media for entertainment and to socialize with friends. Recording TikTok videos is an activity often mentioned by the children. Digital games are also played for entertaining purposes, by adults and children alike, mostly by male participants. Women seem to use more traditional media like books and magazines. Male asylum seekers are more likely to be the ones searching for education online or working while the women look after the children.

In order to help asylum seekers with their information needs, some initiatives were started by different organizations but many asylum seekers seem to rather rely on sources they trust and people from their personal network.

Limitations

Regarding the study's approach, some limitations need to be mentioned. Beginning with the interview rounds, as the interview served as a first starting point to investigate the information behavior of asylum seekers in Germany, some aspects or used services were discovered by the researchers during the interviews. Some aspects were later excluded as they were not seen as important enough. Limitations regarding special groups of study participants, in this case asylum seekers, are also relevant. Even though the language skills of the participants were sufficient and an Arabic translator was present for some of the interviews, not all of the participants spoke Arabic and some misunderstandings could have been possible. Furthermore, the researchers noticed that some participants were very careful from the beginning of the interviews and sometimes hesitated when answering the questions. Concerning the results, age- and gender-dependent differences were observed during the literature review. For example, the older the participants are, the least likely they are to use the Internet. As the study's participants were rather young (the oldest being 55 years, but most of the adults were around 30 years), those differences could not be investigated. Furthermore, the researchers were not able to reach asylum seekers who do not participate in the mandatory language courses. Additionally, even though mainly asylum seekers living in North Rhine-Westphalia were interviewed, as this was the

Federal State with the highest percentage of applications, the results could have been different for participants from other Federal States. As a special consideration, interviewing children presents additional challenges, for example short attention spans. This was also relevant during the interviews as some of the children wanted to play with each other rather than answering questions. Here, a different approach might have been more suitable.

Although the survey was available as an online version, only asylum seekers from integration courses and language classes took part in it. What about those who do not attend language classes or integration courses? At first, the online survey was distributed through online channels like Facebook groups, forums, and survey platforms. Since this was not successful and no one from the target group could be reached to participate in the survey, the researchers visited integration courses and languages classes. Even with this method, only 69 participants answered the survey, which is a very low number in order to be able to make reliable predictions and to present statistically meaningful results. Furthermore, all survey participants are from North Rhine-Westphalia and results of the survey could be different if refugees and asylum seekers staying in other Federal States took part in the study. A difficulty for the survey attendees was the language barrier. Many refugees and asylum seekers and therefore participants of the survey are from Middle Eastern countries and most of them speak Arabic, Farsi, Turkish, and Kurdish. Although the survey was distributed in German language classes with a B1 level to C1 level certification, they still have language insecurities and may not have understood all survey items. Since the researchers were present during the process of filling in the questionnaires, participants were able to ask questions and therefore some uncertainties could be clarified. At the beginning of the survey, the participants who completed the online version were asked to select these social media services, online and traditional media, and ICT devices from predefined lists which they had used in their home country or which they use in Germany. Because of the variety of options, the participants may have missed choosing a specific option. If they missed to name a specific device, media, or service they have used or use, later the participants were not asked about the usage frequency or their motivation to use the service, media, or device. And furthermore, the list did not contain all possible services, media, and devices that were used by asylum seekers in their former home country or are used by them in Germany. The survey questions cover only a small part of the available social media services, online and traditional media, and ICT devices. The survey consisted of pre-formulated options as responses for motives to use a service, device, or media. Those motives given as response items were based on the Uses and Gratification Theory, but attendees were not asked for different or additional motives, if none of the motives applied. Depending on how many services, media, and devices were selected in the beginning, the survey consisted of many questions that needed to be answered. It might be that some participants had to process more than 50 survey items, resulting in possible survey fatigue.

In order to get an overview of the information needs of asylum seekers, a content analysis on questions asked by asylum seekers in a forum called Wefugees was conducted. Although the website is mainly recommended to be used by refugees, everyone is allowed to publish and ask questions on the website; therefore questions which are not asked by refugees or asylum seekers may be included in the evaluation. The website already provides predefined categories for questions, but provides a category "Other questions" as well which serves as the data basis analyzed in this work. Meanwhile, there are more than 2,500 questions asked in the forum; the analyzed data represents a rather small sample with a number of 139 items. It would be more accurate to analyze all questions with the help of an automated content analysis. Furthermore, the numbers of questions assigned to each category on the website are consulted as a comparison. As the asked questions can only be assigned to one of the preformulated categories, intersections of questions between categories are not shown in the numbers provided by the website and are therefore not displayed in the numbers shown in this study's results. Also, the questions were analyzed intellectually by the two researchers, which can lead to a subjective influence on the data. Despite Krippendorff's alpha being calculated and resulting in a perfect intercoder reliability of $\alpha = 0.973$, subjective assessments cannot be completely avoided and eliminated.

Concerning the expert interviews, albeit a small number of interviews are generally a valid scientific approach, by only asking four experts or expert teams (of which one was a group of two) only a small glimpse of the broad topic could be expressed. Furthermore, due to technical difficulties, one recording of the interviews is missing and the information could only be reproduced through notes taken during the interview. As the topic of asylum seekers is quite sensitive, it is possible that some of the answers might have been biased. Furthermore, it is always possible that the interviewers constructed the questionnaire in a certain way which could be perceived as biased or stir the conversation in certain directions.

Outlook

For future studies, taking a look at the information behavior and media practices of the older generation of asylum seekers would be interesting. Especially as this user group is often not as tech-savvy and should therefore not be overlooked. Concerning the questionnaire and interviews, other aspects could be investigated, for example, how the media usage of asylum seekers differs or is similar to the one showcased by German natives or migrants who were not forced to leave their country but rather migrated on purpose. With the knowledge gained from the results of the conducted survey, there is an opportunity to perform an additional survey focusing especially on those services that are used frequently by asylum seekers and refugees. With less survey items, it might be easier to get a more representative number of participants. This would also make it possible to run statistical tests or analyze clusters for further insights. Follow up studies would be equally as interesting to see how the information behavior has changed. However, as all participants were anonymous this is not possible. Therefore, interviewing refugees living in Germany for a longer amount of time could be an alternative.

After arriving at the presented results based on the content analysis, an automated content and sentiment analysis of the forum Wefugees or similar forums and possibly Facebook groups and pages could provide interesting new insights. Especially because of the great need for information and the high number of questions regarding legal aspects, the asylum application, and documents, a more detailed analysis on these topics would clarify where exactly more support and assistance is needed. Also, data from other initiatives and applications who support asylees (e.g., Mbeon) should be collected and evaluated to gain further insights about needed information.

References

Ahmad, M. (2020). A data analysis investigation of smart phone and social media use by Syrian refugees. *Journal of Information and Knowledge Management*, *19*(1), 1–18.

Alencar, A., & Tsagkroni, V. (2019). Prospects of refugee integration in the netherlands: Social capital, information practices and digital media. *Media and Communication*, *7*(2), 184–194.

Bletscher, C. G. (2020). Communication technology and social integration: Access and use of communication technologies among Floridian resettled refugees. *Journal of International Migration and Integration*, *21*(2), 431–451.

Coles-Kemp, L., Jensen, R. B., & Talhouk, R. (2018). In a new land: Mobile phones, amplified pressures and reduced capabilities. In R. Mandryk & M. Hancock (Eds.), *CHI '18: Proceedings of the 2018 CHI Conference on Human Factors in Computing Systems* (Article 584). Association for Computing Machinery.

Emmer, M., Richter, C. & Kunst, M. (2016). *Flucht 2.0. Mediennutzung durch Flüchtlinge vor, während und nach der Flucht* [Escape 2.0. Media usage by refugees before, during and after their escape]. Freie Universität Berlin. https://www.polsoz.fu-berlin.de/kommwiss/arbeitsstellen/internationale_kommunikation/Media/Flucht-2_0.pdf

Katz, E., Blumler, J. G., & Gurevitch, M. (1973). Uses and gratifications research. *The public opinion quarterly*, *37*(4), 509–523.
Köster, A., Bergert, C., & Gundlach, J. (2018). Information as a life vest: Understanding the role of social networking sites for the social inclusion of Syrian refugees. In J. Pries-Heje, S. Ram, & M. Rosemann (Eds.), *Proceedings of the International Conference on Information Systems – Bridging the Internet of People, Data, and Things* (pp. 1–9). Association for Information Systems.
Leurs, K. (2017). Communication rights from the margins: Politicising young refugees' smartphone pocket archives. *International Communication Gazette*, *79*(6–7).
Lloyd, A., Kennan, M. A., Thompson, K., & Qayyum, A. (2013). Connecting with new information landscapes: Information literacy practices of refugees. *Journal of Documentation*, *69*(1), 121–144.
Marlowe, J. (2020). Refugee resettlement, social media and the social organization of difference. *Global Networks*, *20*(2), 274–291.
Mbeon. (n.d.). *About the mbeon project.* https://www.mbeon.de/en/about-the-project/
McCaffrey, K. T., & Taha, M. C. (2019). Rethinking the digital divide: Smartphones as translanguaging tools among Middle Eastern refugees in New Jersey. *Annals of Anthropological Practice*, *43*(2), 26–38.
McQuail, D. (1983). *Mass Communication Theory*. Sage Publications.
Merisalo, M., & Jauhiainen, J. S. (2020). Digital divides among asylum-related migrants: Comparing internet use and smartphone ownership. *Tijdschrift voor Economische en Sociale Geografie*, *111*(5), 689–704.
Merz, A. B., Seone, M., Seeber, I., & Maier, R. (2018). A hurricane lamp in a dark night: Exploring smartphone use for acculturation by refugees. In T. Köhler, E. Schoop, & N. Kahnwald (Eds.), *Gemeinschaften in neuen Medien. Forschung zu Wissensgemeinschaften in Wissenschaft, Wirtschaft, Bildung und öffentlicher Verwaltung* (pp. 308–319). TUDpress.
Mikal, J. P., & Woodfield, B. (2015). Refugees, post-migration stress, and internet use: A qualitative analysis of intercultural adjustment and internet use among Iraqi and Sudanese refugees to the United States. *Qualitative Health Research*, *25*(10), 1319–1333.
Savolainen, R. (2007). Information behavior and information practice: Reviewing the "Umbrella Concepts" of information-seeking studies. *The Library Quarterly: Information, Community, Policy*, *77*(2), 109–132.
Schreieck, M., Zitzelsberger, J., Siepe, S., Wiesche, M., & Krcmar, H. (2017). Supporting refugees in everyday life–intercultural design evaluation of an application for local information. In R. Alinda & P. S. Ling (Eds.), *PACIS'17: Proceedings of the 21st Pacific Asia Conference on Information Systems* (Article 149). Association for Information Systems.
Shao, G. (2009). Understanding the appeal of user-generated media: a uses and gratification perspective. *Internet Research*, *19*(1), 7–25.
Sonnenwald, D. H. (2005). Information horizons. In K. E. Fisher, S. Erdelez, & L. (E. F.) McKechnie (Eds.), *Theories of information behavior* (pp. 191–197). Information Today.
Spink, A., & Cole, C. (2004). A human information behavior approach to the philosophy of information. *Library Trends*, *52*(3), 373–380.
Tirosh, N., & Schejter, A. (2017). "Information is like your daily bread": The role of media and telecommunications in the life of refugees in Israel. *Hagira – Israel Journal of Migration*, *7*, 1–25.
Wilson, T. D. (2000). Human information behavior. *Informing Science*, *3*(2), 49–56.

Glossary

1951 Refugee Convention The Geneva refugee convention is the central legal document of the international refugee law. It contains an official definition of the term refugee and is the basis for the office of the United Nations High Commissioner for Refugees.

Al Jazeera An Arab news broadcaster based in Doha, Qatar, which began broadcasting on November 1, 1996.

Al-Arabiya An Arabic-language news channel based in Dubai Media City in the United Arab Emirates. It was founded on March 3, 2003 as part of the media group MBC.

AnkER facility The central element of the AnkER concept is the bundling of all functions and responsibilities related to the asylum process. All actors directly involved in the asylum process are represented on site in the AnkER facilities.

Asylbewerberleistungsgesetz See Asylum Seekers' Benefits Act

asylee See asylum seeker

Asylgesetz See German asylum act

asylum applicant See asylum seeker

asylum application After registering as an asylum applicant, an asylum application can be filed. It has been expressed in writing, orally or in any other way that one is seeking protection in another country.

asylum law The right of asylum for politically persecuted people is a fundamental right anchored in the Basic Law in Germany

asylum process The requirements for granting international protection (refugee protection and subsidiary protection), the existence of political persecution within the meaning of the Basic Law and deportation bans are determined.

asylum seeker A person who seeks and applies for asylum in a country other than their country of origin, i.e. reception and protection from political, religious, or other persecution.

Asylum Seekers' Benefits Act Determines the amount and form of social or medical benefits that asylum seekers will receive.

asylum status When asylum is granted to an individual, one will receive a specific form of protection (refugee protection, entitlement to asylum, subsidiary protection, ban on deportation).

balkan route Describes the escape routes that refugees and migrants use to get to Europe via the Balkans from the Middle East.

ban on deportation If the general three forms of protection do not apply, a deportation ban can be issued for specific reasons. The applicant will receive a residence permit for one year.

BBC British Broadcasting Corporation is a UK public broadcaster that operates a variety of radio, television, and Internet news services.

Bundesamt für Migration und Flüchtlinge See Federal Office for Migration and Refugees

https://doi.org/10.1515/9783110672022-014

Central Register of Foreigners The Central Register of Foreigners is a database operated by the Federal Office of Administration in which personal data records from foreigners are stored.

CNN Cable News Network is an American television station based in Atlanta, Georgia. It was founded as the world's first pure news broadcaster and began broadcasting on June 1, 1980.

content analysis Content analysis is a bundle of methods from the empirical social sciences. The object is the analysis of the content of communication, which is available in the form of texts, for example as written interviews or newspaper articles. Further studies also focus on audio and video material.

country of usual residence The country of usual residence is the country in which a person lives, the country in which he or she has a place to live.

COVID-19 is a disease caused by a virus named SARS-Cov-2 that led to a worldwide pandemic in 2020.

dating app is an online dating application. It uses GPS data to locate nearby users who can decide based on a profile picture whether they want to date someone or not. If both users agree, they have a match and can exchange chat messages.

desk research Research method using existing (external) data. It is extremely facilitated through the use of the Internet.

digital age also referred to as the information age or computer age, represents the third epoch of social forms after the agrarian society and the industrial age. Encompassing the notion that things are increasingly accomplished by using computers and digital technology. Large amounts of information are available.

digital literacy skills A set of competencies necessary to function in a digital world. Helps to act appropriately and ethically in an online environment. A person's ability to find, evaluate, and clearly communicate information through media and various digital platforms.

digital media Electronic media that are digitally encoded. In contrast to this, there are analogue media. The term is also used as a synonym for "new media."

Dublin III Regulation Criteria for examining an application for international protection to be applied when determining the responsible Member State.

Dublin procedure The procedure is intended to manage and limit secondary migration within Europe. Each asylum application which is lodged in the Member State is examined under the law of one state only.

Dublin rule See Dublin procedure

Duldung is the temporary suspension on deportation of individuals who are legally obliged to leave the country.

EASY quota The distribution quota is determined annually by the Federal State commission and determines what proportion of asylum seekers each state accepts.

economic migrant A person who migrates to another country for economic reasons.

encrypted messaging services Messaging services with security measures (encrypting text and media) to prevent others from reading messages.

encyclopedia An encyclopedia is a particularly extensive reference work encompassing many different areas of knowledge.

entitlement to asylum A person will be entitled to asylum if one is at risk of a serious human rights violation in the event of their return to their country of origin.

EU Common Basic Principles is a set of 11 guidelines by the EU integration policy that should assist EU Member States in the process of formulating their integration policies.

EURODAC A fingerprint identification system for the comparison of the fingerprint data of all asylum seekers older than 14 years.

European Convention for the Protection of Human Rights and Fundamental Freedoms A catalog of fundamental rights and human rights. The European Court of Human Rights in Strasbourg monitors their implementation.

European Dactyloscopy system See EURODAC

European migrant crisis The European migrant crisis is understood as the sharp increase in the number of asylum seekers in EU countries in 2015.

European Migration Network A network that is financed by the European Union with the aim of providing EU organs, national institutions, and authorities as well as the general public with up-to-date, objective, and comparable data on migration and asylum.

Eurostat The Statistical Office of the European Union is the administrative unit of the European Union for the compilation of official European statistics.

expert interview A research method used to gain insights of experiences or knowledge of specific and qualified personnel.

face-to-face interviews Interviews that are conducted by trained interviewers using a standardized interview protocol. They offer the advantage to control interactions and ask complex questions.

Facebook A social network service connecting millions of people worldwide.

Facebook Messenger is a messenger from the Facebook social network service which allows users to exchange text messages and other media files.

fake news A term used to refer to manipulatively distributed, simulated messages and news that are mainly spread virally on the Internet.

family asylum applies to members of one family. If a person has been recognized as entitled to asylum, their family members residing in Germany can also apply for asylum.

family reunification To protect marriage and family, spouses or registered partners, parents, and unmarried underage children of individuals to whom entitlement to asylum or refugee status has been granted have the opportunity to move to Germany with their relatives.

Federal Office for Migration and Refugees (Bundesamt für Migration und Flüchtlinge) The Federal Office for Migration and Refugees (BAMF) is the competence center for asylum, migration, and integration in Germany.

Federal State (Bundesland) is one partially sovereign member state of the Federal Republic of Germany.

first-time asylum application An asylum application that was filed by an individual who applies for asylum for the first time.

forced migrant A person forced to migrate due to threats to life and livelihood. Movements of refugees as well as people displaced by natural or environmental disasters.

forced migration See forced migrant.

foreigners' registration office Office responsible for issuing or refusing residence permits according to the respective purposes of residence, decision on the issuing of settlement permits, the decision and implementation of deportations.

form of protection Asylum applicants can get granted three different forms of protection: refugee, entitlement to asylum, and subsidiary protection.

forum A forum in the general sense is a real or virtual space where questions are asked and answered or content about a specific topic is exchanged.

Fox News Fox News Channel (abbreviated FNC, often just called Fox News or Fox) is an American news channel based in New York. It went on the air on October 7, 1996.

Geneva Refugee Convention see 1951 Refugee Convention

geolocalisation The determination of the location in relation to a defined fixed point.

German asylum act (Asylgesetz) Law determining the asylum procedure in Germany.

German asylum procedure Individual steps necessary to determine the form of protection of a person.

German Federal Foreign Office Office responsible for the country's foreign policy and its relationship with the European Union.

Google Search engine to find documents on the Internet.

Google Maps Online map service displaying locations of institutions or known objects, view of roads, and the location of oneself. Further allows route planning and navigation functions.

Google Translate An online service that machine translates words, texts, and entire web pages into another language.

GPS The Global Positioning System (GPS) is a global navigation satellite system for determining positions.

gratifications obtained approach based on the Uses and Gratifications theory adding new insights. A person's needs are fulfilled by media consumption.

gratifications sought approach based on the Uses and Gratifications theory adding new insights. What a person expects to experience from media usage.

Handbook and Guidelines on Procedures and Criteria for Determining Refugee Status Handbook under the 1951 Convention and the 1967 Protocol Relating to the Status of Refugees.

health information Information related to health issues and health data.

home country Country in which one was born or raised and which is one's country of usual residence.

homeland See home country.

host society Residents of a country in which someone applies for asylum in.

ICT skills Skills and manners associated with the adept handling of ICT.

ICTs Abbreviation for Information and Communications Technology. See information and communications technology.

immigrant Person who immigrated to a country. From the perspective of their country of origin, migrants are emigrants; from the perspective of the host country, they are immigrants.

immigration The process of moving to and adapting to a new society of a host country.

information and communications technology is a term that describes technology utilized in the area of information and communication. It includes communication applications, radio, television, mobile phones, smart phones, hardware and software for computers and networks, satellite systems, and the various services and applications associated with them.

information behavior A discipline of information science research concerned with the understanding of how people produce, search, and use information.

information broker People (or institutions) who research information.

information campaign Aimed at driving awareness to different topics.

information channels Channels used to provide information. Can also describe types of physical transmission mediums.

information deficit is present if a person experiences a knowledge gap. Further it can also be seen as a divide between people who have the needed information and people who do not.

information dissemination is a method to distribute information among individuals and also to provide information to the general public.

information gap Refers to an activity in which a learner is missing some crucial information (to complete a task).

information horizon Concept describing a person's knowledge. In every context and situation is an information horizon present in which someone can act. It is consistent with a variety of information sources and determined by social and individual factors.

information literacy The ability to deal with information in a self-determined, confident, and goal-oriented way.

information need is a lack of (relevant) information experienced by an individual and the desire to locate and find this information.

information needs matrix A model derived from refugee information behavior investigations and displays information needs of refugees during integration. It was developed by Oduntan and Ruthven (2019). "The study found information needs in refugee integration centered on the host society's provisions for sociological needs such as housing, support, legal needs, mobility, health, education, social, employment and benefits."

information practice are methods and procedures applied to receive and distribute information.

information precarity A situation in which a person is missing crucial information that could lead to several consequences.

information production is the process of creating and publishing content as a document.

information reception is the process of consuming knowledge.

information seeking is the process of searching for information and knowledge, e.g., on the Internet, via databases, or face-to-face.

information service A service used to find information with the help of an information retrieval system.

information technology is a term to describe electronic data processing and the used hardware and software infrastructure.

Instagram A picture and video-sharing social networking platform on which users can upload media and other users can comment on and like the uploaded media.

integration is a procedure during which newly arriving people get involved into and introduced to the society of a new country.

integration initiatives Organizations and initiatives which support newly arriving people to get involved with and introduced to the society of a new country.

integration procedure see integration

International Organization for Migration (IOM) A intergovernmental and transnational legal organization in the UN system that operates assistance programs for migrants.

interview A research method used to collect data by asking research focus groups several questions regarding a specific topic.

intuitive research model is a model that is derived from theory in order to display a phenomenon or process.

irregular migration is a form of illegal immigration, e.g., someone with a visa does not leave the country after the visa expired.

Islamic State of Iraq and the Levant A radical Islamic militant group; also known as Islamic State (IS) and Islamic State of Iran and Syria (ISIS).

Königsteiner Schlüssel is a regulatory in Germany which defines the financing share of the federal states regarding the distribution of asylum seekers based on the federal state's population and tax revenue.

language barrier A phrase used to describe difficulties in communication experienced by people who speak another language.

language brokering is a phenomenon where youth translate the language for their parents or other adults.

language level A guideline to describe the experience grade of language learners.

Lasswell formula is a sender-centered model in communication research named after Harold Lasswell who investigated the model in 1948. He formulated a sentence consisting of five questions: “Who? Says What? In Which Channel? To Whom? With What Effect?” (Lasswell, 1948, p. 37)

Likert-scale A scale used in questionnaires to represent the attitude of research participants toward a particular topic.

LinkedIn LinkedIn, based in Sunnyvale, California, USA is a web-based social network for maintaining existing business contacts and establishing new business connections.

literature review A method used to retrieve and summarize research sources on a particular topic.

live streaming services are a kind of social networking service on which streamers can broadcast live videos while interacting with an audience. The audience may chat with the streamer via chat messages.

manifestly unfounded is a decision on an asylum application in which the application was not approved. The individual is granted a time period of 7 days to leave the country.

mbeon is a mobile advice-giving application for refugees and migrants in Germany. Via messaging-service people can ask questions and ask for information.

messaging service is a form of online chat that allows sending instant text messages and media between users.

micro-blogging service A form of short-message online blogging platform.

Middle East is a geographical region including countries from the Southern and Eastern Mediterranean Sea (Turkey, Cyprus, Egypt, Libya), the Arabian Peninsula, the Levant, Iran, and Iraq.

migrant Describes a person who changes his or her place of usual residence, usually migrating to another country.

migration The act of changing one's place of usual residence.

migration flow is the number of migrants who cross a border within a certain time period.

migration infrastructure are technologies, initiatives, actors, and institutions that interact and are linked to promote migration.

MILo (Migration Infologistics) An international information system of the Federal Office for Migration and Refugees to retrieve information about migrants and refugees .

misinformation Misleading information which may contain inaccurate facts.

multiple methods approach is an approach whereby multiple qualitative or quantitative methods are applied to collect data for answering one research question or several research questions.

new media see digital media

news channel A news channel is a special interest program on television or radio with a focus on news. It is usually broadcast around the clock, with the program containing news programs at least every half hour.

news website are online websites of news agencies and newspapers.

non-government organizations (NGOs) are organizations that are independent from the government.

official websites Marked websites from, e.g., the European Union, all official websites are marked with the domain europa.eu.

old home country see home country

old media are all kinds of media that are not connected to the Internet. E.g., radio, television, print media.

opinion leader A person who shows a high degree of knowledge in a particular matter of public and political interest.

outright rejection is a decision on an asylum application in which the application was not approved. The individual is granted a time period of 30 days to leave the country.

permission to reside is a certificate one receives after filling an asylum application. The document proves that a person resides in a European country lawfully. But the persons are not permitted to work and only allowed to temporarily leave the area they are obligated in.

personal interview Part of the asylum procedure (Dublin procedure) to decide whether one's asylum application will be granted or not. Asylum applicants will have to attend an appointment where they have to state their biography and situations as well as reasons for taking flight. A decision-maker from the Federal State will be responsible for holding the interview.

persons entitled to protection are individuals who have received an entitlement to asylum, refugee protection or subsidiary protection in an European country.

persons entitled to remain are individuals who are legally allowed to stay in a European foreign country but did not receive a form of protection.

Pinterest Pinterest is an online pin board for graphics and photographs with an optional social network including a visual search engine.

podcast are audio reports or series which are available for download or broadcasting on the Internet.

PRISMA flow diagram A diagram used to represent the flow of information through the various phases of a systematic (literature) review.

proof of arrival is a document that proves the registration of an asylum applicant at a reception facility or arrival center.

refugee is a person whose asylum application has been granted and is given refugee protection in accordance with the Geneva Refugee Convention.

refugee protection is granted to individuals who apply for asylum in a country. Based on the basic law for the Federal Republic of Germany, the refugee status is limited to politically persecuted persons.

refugee status see refugee protection

Refugee Status Determination (RSD) An administrative or legal process by governments or UNHCR to determine whether a person seeking asylum is considered a refugee under regional, national law.

Refugee Status Determination process see Refugee Status Determination (RSD)

Residence Act The legal basis for the entry, leaving, and residence of foreigners in Germany.

residence permit Three different types of German residence permit exits: Temporary residence permit; permanent residence permit; the blue card.

retrieval system A platform to search for and to retrieve information and documents.

rumor is an unconfirmed statement which is told among people that might not be true or only partially be true.

Scopus An online citation and literature database and information retrieval platform.

search engine is a web based retrieval system.

search query is a text containing words, phrases, and optionally Boolean operators that is typed into a search engine in order to retrieve information and documents from a database.

Skype An Internet based VoIP-telecommunication and video call application with instant-messaging chat function.

SMS is the abbreviation of short message service. It is a service that allows mobile devices (usually mobile phones) to exchange brief messages.

smuggler is a person who illegally transports other people for a fee and usually takes them to another country or to a certain place.

Snapchat A free-to-use instant messaging service which allows users to send media (mostly photos and videos) to other users that are only visible for a certain period of time (usually a few seconds) before they are automatically deleted.

social media are digital online media services which enable their users to connect with other users and upload, publish, and share media content.

social networking services (SNSs) is an online platform on which users connect with other users and form networks or communities.

stopover country A country in which asylum seekers settle for an uncertain time in order to rest, get information, and plan further steps for the escape route.

streaming service A service offering video on demand or audio files to enjoy at the user's chosen time.

subsidiary protection is a form of protection someone who is seeking for asylum receives in an European country and is not legally seen as a refugee.

survey A research method to ask a representative population questions and collect data about certain topics.

Taliban is a radical Islamic military organization and political movement in Afghanistan.

target country is the country in which an asylum seeker stays and one usually applies for asylum in.

Telegram is a mobile instant messaging service which allows users to send end-to-end encrypted messages, or other media including photos, videos, and files. It also allows users to take video and audio calls.

temporary suspension of deportation see Duldung

The Migration Advice Service for Adult Immigrants (MBE) is an integration initiative and advisory service for migrants older than 27 in Germany; in German: Migrationsberatung für erwachsene Zuwanderer.

theoretical framework is used to explain phenomena in research studies by combining different theories.

TikTok TikTok is a video portal for lip-syncing music videos and other short video clips, which also offers functions of a social network and is operated by the Chinese company ByteDance.

traditional media see old media

translation service A digital service used to translate documents instantly. It can also offer spoken translations.

Twitter is an online micro-blogging and social networking service where users can post, comment, like, and repost messages, so called "tweets."

unaccompanied minors are children and adolescent asylum seekers younger than 18 years without being accompanied by a parent or any other adult being responsible for them.

United Nations High Commissioner for Refugees (UNHCR) is a UN agency whose duty and mission is to protect refugees and support refugees and their rights.

Uses and Gratifications Approach see Uses and Gratifications Theory

Uses and Gratifications Theory is an audience-centered approach investigated by Katz, Blumler, and Gurevitch (1973) to understand why and how people actively use media to satisfy their needs and further, which needs are satisfied by media consumption.

Viber A mobile instant messaging service which allows users to send end-to-end encrypted messages, or other media including photos, videos, and files. It also allows users to take video and audio calls.

virtual private network (VPN) service is a private safe network connection to another server or database from a computer via the Internet that provides security protection by encrypting IP addresses.

visa is an entry permit to a foreign country confirmed by authorities. Its validity is granted for a certain period of time.

Web of Science An online citation and literature database and information retrieval platform from Thomson Reuters (also formerly known as Web of Knowledge).

Wefugees is an online forum in which refugees can ask questions. The questions will be answered by volunteers. It should serve as a first step to live an independent life in a new country.

WhatsApp is a mobile instant messaging service which allows users to send end-to-end encrypted messages, or other media including photos, videos, and files. It also allows users to take video and audio calls.

WickR A mobile instant messaging service which allows users to send end-to-end encrypted messages, or other media including photos, videos, and files.

Wikipedia is a free-to-use online encyclopedia. It is a crowdsourcing project, which is available in many different languages.

Yahoo! Messenger An online instant messaging service which was, until 2018, free-to-use with a Yahoo! account.

YouTube is a free online video-streaming platform where users can upload their own videos, comment, and vote on uploaded content and broadcast live streams.

Name index

https://doi.org/10.1515/9783110672022-015

Subject index

https://doi.org/10.1515/9783110672022-016